KT-549-763

How Much Brain Do We Really Need?

Alexis Willett, PhD and
Jennifer Barnett, PhD

ROBINSON

First published in Great Britain in 2017 by Robinson

Copyright © Alexis Willett and Jennifer Barnett, 2017

1 3 5 7 9 10 8 6 4 2

The moral rights of the authors have been asserted.

All rights reserved.
No part of this publication may be reproduced, stored in a retrieval system, or transmitted, in any form, or by any means, without the prior permission in writing of the publisher, nor be otherwise circulated in any form of binding or cover other than that in which it is published and without a similar condition including this condition being imposed on the subsequent purchaser.

A CIP catalogue record for this book is available from the British Library

ISBN: 978-1-47213-896-5

Typeset in Adobe Jenson Pro by SX Composing DTP, Rayleigh, Essex
Printed and bound in Great Britain by Clays Ltd, St Ives plc

Papers used by Robinson are from well-managed forests and other responsible sources

Robinson
An imprint of
Little, Brown Book Group
Carmelite House
50 Victoria Embankment
London EC4Y 0DZ

An Hachette UK Company
www.hachette.co.uk

www.littlebrown.co.uk

For Graham, without whom this book would literally not have happened, and for Cassidy, whose hilarious distractions forced me to time-manage my writing like never before! *You're* the best.

A.W.

For Greg, and the weird and wonderful developing brains he has chosen to share with me.

J.B.

Contents

Acknowledgements

This book wouldn't have been possible without the generous help and advice of other people, many of whom know a great deal more about the brain than we do. A particular thank you goes to our expert interviewees, Graham Murray, Laurie Weiss, Simon Kyle, Fergus Gracey and Maggie Alexander, for their time and for agreeing to have their fascinating thoughts included in the book. Huge thanks also to the following wise friends for their great suggestions: Anna Barnes, Alexia Barrable, Nick Walsh, Joel Walmsley, Ross Rounsevell, Lizzie Gross, Henrie Aitken and Juan Sanchez-Loureda.

Any inaccuracies are despite their best efforts, and entirely our own.

Finally, thank you to our wonderful families and friends, who have been terrifically enthusiastic and supportive of this endeavour (and who we hope will be first in the queue to purchase their copies!).

INTRODUCTION

Tackling the Question

Bang. The universe appeared. Then ape-men. Who then evolved. Their brains grew and grew, then shrank a bit, and now here we are. Admittedly, this isn't the most scientific synopsis of how we got here today, and obviously there has been a lot of other stuff going on too, but what we want to know is: why did we grow so much brain and do we really need it all now and in the future?

Before we go any further, why are we even talking about how much brain we need? A chance conversation overheard between two outstanding psychiatrists prompted the titular question tackled in this book. The psychiatrists were discussing the results of a study in which brain scans revealed that some people who had taken antipsychotic medication for schizophrenia experienced brain shrinkage. They wondered what this shrinkage meant in practical terms for these people's day-to-day lives, as this was not recorded within the study. This opened up a debate as to whether the brain shrinkage would have been noticed by the individuals at all, had the brain scans never been taken. In other words, did the brain shrinkage actually matter? Which then led us to ask whether brain shrinkage is something unique to that study. Should the rest of us be concerned about what might be happening to our brains?

In fact, it turns out that brain shrinkage is no rare thing. Whilst the bad news for all of us is that from our mid-thirties our brains start to shrink, the good news is that we generally still seem to muddle along. But how can this be? we wondered. Does this mean we don't really need all of our brain?

We humans all have brains. Admittedly, this might be considered debatable in the case of some reality-TV-show contestants and certain politicians, but nonetheless, brains we all have. The wobbling mass of cells and connections that teeters delicately atop our bodies controls our every move and thought. It is a supremely complex organ, more so than any other in the body, and holds responsibility for everything that we are. Despite huge variation in the human form across populations, healthy, fully developed brains have pretty much the same architecture.

The modern human brain consists of a number of key parts, broadly grouped into the cerebrum, the cerebellum and the brain stem. Each of these parts has many components (we've tried to be helpful by including a basic figure for your reference: Figure 1). The cerebrum is the large bit that we mainly think of when we imagine a brain; looking like a scrunched-up spongy sausage. The cerebellum is much smaller and located at the rear of our head, beneath the cerebrum. In front of the cerebellum, underneath the cerebrum, is the brain stem. However, no doubt you already know that this is a highly simplistic overview of our brain anatomy in its grossest form.

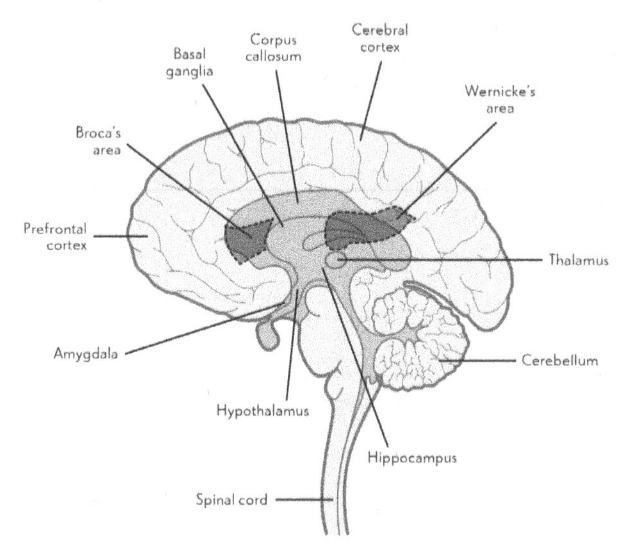

Figure 1: Key Structures of the Brain

Within the context of the entire animal kingdom, the human brain is many times larger than would be expected for our body size. To allow such a large brain, it is argued, human babies must be born earlier than other species, defenceless, unable to feed themselves or even move out of harm's way. From a survival point of view that doesn't seem sensible. So why did evolution shape our brain in such a way, and what advantage does all this extra brain bring?

The brain is the most complex system we know, a piece of hardware that trebles in size over the first year of life and then constantly reshapes and reprograms itself at the microscopic level in response to every new life experience. Some experiences, like education, result in more brain cells and a larger brain. With others, like drinking alcohol and even normal ageing, cells die and cannot be replaced. People with larger brains function better on average, and live longer. Yet there is great variation within our species: men's brains are around 10 per cent bigger than women's, and people whose ancestors developed in colder climates tend to have a larger, rounder head and brain, in order to better conserve heat. The impact of these differences in 'hardware' is mitigated by differences in 'software': peak brain function depends not only on the size of the brain, but on how efficiently it is used.

Contrary to the popular myth, we do use more than 10 per cent of our brains, but even modern neuroscience cannot accurately tell us how much of our brain we use at any one time, or ever. We do know that surprisingly large amounts of brain can be damaged or lost without our losing functions like language, thinking and emotions, which seem to be the core of the human experience. Does this mean that losing a few brain cells here and there would be something we wouldn't even notice?

People are endlessly fascinated with the brain and the mind, whether seeking greater understanding of themselves and those around them, or just marvelling at the brain's intricacy and the many secrets it still seems to hold. Whilst many writers have sought to present the workings of the mind and the brilliance of mental achievement, *How Much Brain Do We Really Need?* challenges you to think differently about the brain. Rather than concentrating on the many wonderful things the brain can do, we ask whether in fact we can live satisfactorily without some of it.

What is in this book anyway and why should you read it?

This book aims to shed light on the capabilities of the human brain – in both optimal and suboptimal conditions, and in the past, present and future – and considers what it may do without.

It presents intriguing hypotheses, compelling individual experiences, and professional points of view. To help you navigate, we've divided the book into four parts, which can be broadly summarised as: Part One – who we are and how we got here; Part Two – normal variation of the brain; Part Three – when things aren't quite as they should be in the brain; and Part Four – where the brain might be heading in our lifetime and beyond.

We shall reflect on human evolution, asking why we have so much brain in the first place and what distinguishes us from other species. We'll then consider how we define what a working brain needs to be able to do, from the basics of survival to the highest levels of human achievement. To do this, we examine evidence from development across the human lifespan, and take a glimpse into the effects of brain injuries, abnormal development and degenerative diseases. We explore what normal variation in brain size and structure between the sexes, and within populations, means for the relationship between brain biology and intelligence and other cognitive functions. And by looking at what is known about normal variation in brain function and assessing it alongside cases of people who live without a wholly functioning brain, we'll ponder whether we need so much brain and whether a little bit of shrinkage does us any harm after all. To round things off, we gaze into our crystal ball and contemplate the future possibilities for our brain, given optimal nutrition, education and medical care, and the potential ability to replace or re-engineer faulty or ageing brain cells.

But you don't just have to listen to us. A host of brain experts also shared their thoughts on how much brain we really need and kindly allowed us to convey them to you in this book. Don't say we don't spoil you. Woven throughout the book you will find interviews with psychologists, psychiatrists, neuroscientists and those who work with people affected by neurological conditions, offering fascinating insights into the brain.

However, we also want you to prepare for a little cognitive exercise and do some of the work. As you read it, we want you to ask yourself repeatedly: How much brain do we really need? We will arm you with facts and figures, case studies and hypothetical scenarios, expert interviews and scientific principles. We'll take you on a journey from the ancient mists of time to the far reaches of the future, via different species and lands. We will tackle diverse questions, such as whether brain-training or being a 'SuperAger' is the key to healthy ageing; whether drinking coffee or going for a run is better for our cognitive performance; whether women really experience 'baby brain'; whether screen time is ruining our brains; and, of course, whether marmosets make good detectives.

A VIEW FROM

Dr Graham Murray, University Lecturer, Department of Psychiatry, University of Cambridge, and Honorary Consultant Psychiatrist at CAMEO Early Psychosis Service, Cambridgeshire and Peterborough NHS Trust, UK

Graham juggles two main roles, being both a researcher and a clinician. He is a neuroscientist and psychiatric researcher with a particular interest in brain imaging. His research focus is on cognitive and brain development over the lifespan, and the physiological basis of mental illnesses. As a psychiatrist, he works in the field of 'early intervention in psychosis', meaning that he specialises in the treatment of young people who have developed schizophrenia and other psychotic illnesses for the first time. Graham was one of the two psychiatrists in the conversation that prompted this book, and he co-led the study in question.

So, Graham, tell us a bit more about your study. What were you looking for and why?

'We were interested in whether there is progressive neurodegeneration in schizophrenia. Psychiatrists have long been interested in this question, because some people with schizophrenia have a declining course to their illness. By this I mean they become increasingly cognitively disabled the longer they are unwell; their symptoms get worse and they seem to have more difficulty with everyday activities that require mental function. Things can deteriorate for some to the extent that they can't live independently and need to live in housing where there is twenty-four-hour support. It is not known why some patients have a deteriorating course, but one long-standing theory is that it might be caused by progressive brain atrophy (shrinkage).

'So we wanted to know whether there is brain volume loss over time in schizophrenia. However, we know that we all lose brain volume as we age, so we needed to measure the rate of brain volume loss in people with schizophrenia compared to that in a control group of those without the condition. We know from our clinical experience that schizophrenia is a very variable condition and, whilst some patients are very disabled by the illness, others make a complete recovery. So we were interested in examining the variability of brain changes in schizophrenia. We specifically looked at whether medication might be associated with either a decrease or increase in the rate of brain loss, and whether patients with greater brain volume loss had symptoms, or cognitive functions, that were getting worse.

'For each person we took two brain scans, at an interval of around nine years. We used a computer algorithm to calculate the degree of brain volume loss between the scans. We found that, on average, those with schizophrenia had a higher rate of brain shrinkage than the controls, and the degree of shrinkage experienced was correlated with the amount of antipsychotic medication they took.'

Were you and your colleagues surprised by this result, or was it something you were expecting?

'It wasn't something we were expecting at the time we started the study, as many researchers had previously claimed that the medication was neuroprotective and prevented brain volume loss. However, the study took several years to conduct and by the time we had our results, another group from Iowa had performed a similar study with the same result. So effectively we replicated their finding.'

Taking all these results in context, what do you think is going on in the brain?

'The simplest explanation is that the medication is causing accelerated brain shrinkage. However, in a purely observational study like ours it is very difficult to prove causality. It is possible that the people with schizophrenia with the most severe illness had the most brain shrinkage, and that in an attempt to help these severely ill patients the treating doctors had resorted to prescribing increasingly high doses of medication. We know that the higher the dose of medication, the greater the loss in brain volume; but we can't say for sure whether the loss was because of the greater amount of medication or whether the people who needed the most medication (i.e. those who were most ill) had brains that were more likely to shrink. In other words, you can't fully tease apart the effect of the medication from the effect of the illness on the brain.'

Do you think brain shrinkage has a noticeable effect on people? Is this something you take into account in your clinical work?

'We found no link between the degree of brain shrinkage and changes in cognitive function. In other words, there was no demonstrable effect on mental performance. What we found in schizophrenia is not the typical picture of classic neurodegenerative diseases like dementia. We think that in dementia the brain loss drives the clinical progression of the illness, but in schizophrenia the picture is much less clear. It is too simplistic to think that brain shrinkage is always a bad sign. In some circumstances, it seems to be helpful that certain parts of the brain are shrinking. For example, in normal development children's brains have less grey matter as they get older (though white matter is increasing), and studies have indicated that the children who lose the most grey matter in early adolescence make the most gains in intellectual function. So thinking back to our study, whilst at first glance it may seem alarming to think that people with schizophrenia had high rates of brain loss, it could possibly be that this loss is a process of the brain adapting somehow, and could even be beneficial.

'So you can probably see by now that using a measure like brain-volume loss in psychiatric clinical work wouldn't currently be particularly helpful, as it is unclear what it all means. Maybe in future we will be able to incorporate such measures to inform patient care when we have a greater understanding of what brain-volume loss is indicating at an individual level.'

How much brain do you think we really need?

'Most of us seem to be able to manage pretty well despite the fact that our brains shrink year on year as we get older, after the age of about thirty. That's not to say that the process is harmless: it could be that eventually we can no longer compensate for the losses, and at that point mental function might suddenly decline. We should also remember that it is not just about the amount of brain we have, but how efficiently the brain works. This may relate to how well connected the different regions are, to allow information to flow efficiently, and to the balance of chemical messengers in the brain that are critical to this. So perhaps the answer is that we ideally need all of our brain, but we typically have the ability to adapt to changes in the brain.'

PART ONE

A Matter of Matter

How much brain do we have and
does it matter?

CHAPTER 1

Head of the Class: Why do humans have so much brain?

Our brain wasn't always a mass of beautiful complexity. Whilst it has evolved phenomenally since we dragged ourselves from the primordial soup, the changes didn't exactly happen overnight. It has taken billions of years to craft this amazing organ, since long before our earliest ancestors walked the planet.

Before we delve into what our brain does and whether we really need it all, we're going to take a step back in time and consider why we have the brains we do. Understanding something about the evolution of the brain helps us to identify what parts may matter most (and least), and why we are the way we are now. From our early ancestors to the present day, the human brain has grown but is now starting to shrink again. How has the human brain evolved, and how do we compare with other species?

Let's start at the very beginning (it's a very good place to start)

As life on Earth began, brains as we know them didn't exist. We all started out as tiny bacteria, with no discernible brain, and existed like this for billions of years. Over time, however, evolution favoured those organisms able to find nutrients and avoid risks; and gradually these primitive creatures began to develop into something more exotic. That 'something' required first the development of a control system capable of carrying out more sophisticated behaviour (not just immediate responses to stimuli, for example) and then, much later, advances that allowed that organism to better coordinate its behaviour with that of others of its species.

The nervous system slowly evolved as some cells (neurons) became specialised for carrying messages, and developed long extensions (axons) to communicate with other cells where they met them, at the synapses. As this nervous system evolved, the brain came into being as its control centre. Groups of neurons came together to form what we call the central nervous system, allowing more complex processing of information and enabling animals to move and respond to the environment in more elaborate ways.

The brain slowly got bigger and better. The oldest bits of the human brain, in evolutionary terms, are those that keep us alive, controlling things such as our breathing, heart rate, body temperature and balance. If you are indeed alive while reading this, you may have guessed that these parts are still in the brain today (more about them in Chapter 2); but then newer, more fancy and sophisticated structures evolved to improve our brain's abilities no end.

Eventually, the capacity to learn and remember developed, and neural processing became increasingly efficient. As brains were exposed to greater visual, auditory and other sensory input, they developed something called a neocortex (which we'll refer to as 'the cortex' for ease from now on). This is the most recent addition to our brain and is often considered to be quite special in terms of human mental prowess. The cortex enables complex activity, particularly social behaviour, and so its emergence paved the way for more intricate movement, conscious thought, judgement and, eventually, language.

As you might imagine, the first mammals to appear on the planet (around 200 million years ago) only had a small cortex. Some of these animals took to the trees, and in order to adapt to their new lifestyle they needed better coordination to navigate their environment and better eyesight to catch fast-moving prey, such as insects. This change in behaviour led to the expansion of the visual part of the cortex, as those individuals best adapted to tree life more successfully passed on their genetic advantage. More complex connections between different parts of the brain became established and mammals, especially primates, were then able to behave in ever more sophisticated ways.

So you can see that brains had come a long way prior to even the emergence of hominids (i.e. all species of extinct and modern great apes, which include us humans). Although modern man's earliest ancestors lived around 6–7 million years ago, a number of hominid species emerged and subsequently disappeared before we came on the scene. Modern humans (*Homo sapiens*) only appeared a mere 200,000 years ago. So what was ape-man doing all this time and, more importantly, what was happening to his brain?

A brief note on evolution

Before taking a closer look at how the hominid brain morphed into the impressive powerhouse we now hold so dear, let's remind ourselves how evolution is supposed to work. Charles Darwin kindly applied his own notable brain to give us the theory of evolution. In his 1859 tome *On the Origin of Species,* Darwin postulated evolution by natural selection, a process whereby organisms alter over time due to changes in physical or behavioural traits that are inherited over generations. Beneficial changes that enable an organism to better adapt to, and thrive in, its environment will help its chances of both survival and reproduction. Numerous different species in the animal kingdom evolved over time, becoming more sophisticated in their abilities and refined in their physical form. It is worth noting, however, that evolution isn't necessarily linear in its pattern. Changes can occur at any time and in any way, and those that prove to be advantageous might branch off in another direction, creating something new rather than just a refinement of the old. For example, although we know we are closely related to other primates and often vainly consider ourselves a more sophisticated version of them, it isn't the case that modern humans evolved from monkeys. As we know, there are many species of monkey alive and well today; both modern monkeys and modern humans evolved from a common ancestor along distinct branches of change.

In order to gauge how the human brain evolved from those of our earliest ancestors, we need some evidence. Unfortunately, the world is not overflowing with prehistoric brains that we can dissect, scan and interrogate to determine how the human brain became the way it is now. Brains are excellent at many things, but fossilising is not one of them. However, their protectors, the ever-faithful skulls, are much better at this and lend themselves nicely to researchers evaluating their size and shape to estimate changes that have taken place over time. Ancient artefacts also help to build a picture of how early humans lived, and these can help us to infer something about the abilities their brains must have had. We can also try to infer some changes that were taking place by our understanding of what functions and abilities would have been needed to survive, thrive and behave in the way we think our early ancestors did. It is important to note that many theories about brain evolution based on evidence from the fossil record are up for debate, and new ones continue to emerge. Despite being unable to determine the exact evolutionary path, when we study the cranial vaults of different hominid species it is undeniable that our brain has been increasing in size and changing in shape over time.

What were our brains like back in the mists of time?

Have our brains really changed all that much since our ancestors started walking upright? One of the earliest species (around 6–7 million years old) identified as part of the human family tree is *Sahelanthropus tchadensis*. The fossil remains of one individual were only relatively recently discovered, in 2001. Because only fragments of skull were found, its size can only be estimated, but it is thought that this species had a skull, and presumably a brain, slightly smaller than that of a modern chimpanzee. To put this into context, chimpanzees have brains that are around three and a half times smaller than those of modern humans. OK, we know that's only one fossil, but it does seem to indicate that the brain has grown substantially over time. However, to press the case we probably need a bit more detail on what's been happening over the last few million years.

Later in evolution, species in a group termed *Australopithecus* developed an advantageous combination of features of both apes and modern humans. Whilst they were bipedal and regularly walked upright, they also had long arms with curved fingers that enabled them to skilfully climb trees. It is thought that such adaptations helped this group to survive and thrive, since its members were able to live in both trees and on the ground as the climate and their environments changed. A well-known personality from this group is 'Lucy', from the *Australopithecus afarensis* species, dated to around 3.2 million years ago.

As she sat in a tree in what is now Ethiopia, contemplating where to get her next meal, Lucy could never have imagined that one day she would become arguably the world's most famous member of her species; indeed, one of the most famous of all fossils. At the time of her discovery in 1974, Lucy revolutionised our understanding of human origins. Not only was she one of the oldest known fossils at the time, she was also one of the most complete, which helped to capture the public's imagination. Seeing her skeleton, people could imagine and marvel at the existence of a small, human-like creature sharing the same planet as them but millions of years before their time.

Being able to navigate both ground and trees must have meant an alteration in Lucy's brain and those of her contemporaries. The information being processed by the brain and the instructions it would have given the body were presumably more complex than those experienced by earlier species, and this would require an upgrade in mental equipment. Initially, though our early ancestors' brains were within the size range of those of other apes living today, there were subtle differences in brain architecture in comparison with apes as the cortex began to expand, suggesting development in higher functions.

Whilst species from the next branch on our family tree, the *Paranthropus* genus, display a further small change in brain size, it is not until the emergence of the *Homo* genus that major transformations in the brain can really be detected.

The distinction of the *Homo* genus, of which the modern human is just the most recent species, was defined in 1955 as an upright posture, a bipedal gait and the dexterity to fashion stone tools. The new kid on the block, the first of the genus, was *Homo habilis,* known as 'the handyman' due to its supposed tool-making skills. This species, dated at 1.4–2.4 million years old, had a braincase (of 600 cubic centimetres) half the size of a modern human brain. This was around 50 per cent larger than any of the *Australopithecus* gang. Some researchers have considered *Homo habilis* to just be a variation of the next species of hominid, *Homo erectus,* rather than a distinct species, but either way, observable changes in the brains of this new genus were taking place. Examples of *Homo habilis* fossils show they had developed a slightly bigger brain, which included expansion of a part involved with language called Broca's area. This suggests they were communicating with each other, although we don't know whether language as we might recognise it had developed.

Bona fide examples of *Homo erectus* had a bigger brain still. The brain size of *Homo erectus* is estimated to have been around 900 cubic centimetres, significantly larger than that of its predecessor *Homo habilis* but still a way short of the modern human brain.

Being able to make and utilise tools enabled our ancestors to benefit from a more varied and energy-rich diet, and once fire was harnessed to cook meat they were able to obtain even more nutrients. It has been demonstrated that cooked meat has a higher nutritional value than raw meat and, as cooked meat is easier to chew and digest, you need less of it to gain the same amount of energy. This meant that our early ancestors no longer needed such large guts (which used a lot of energy in processing food) and over time this part of the anatomy got smaller, freeing up more energy for brain expansion. Modern humans have a very small gut in comparison with other great apes (about 60 per cent of the volume). However, our bigger brains take a lot more energy to power, meaning that our diet needs to be more nutrient-dense to achieve the fuel required to function effectively. Primatologist Richard Wrangham at Harvard University believes that the point at which our bodies show adaptation to cooking is 1.9 million years ago, which he suggests was a key watershed in our divergence from other primates.

As modern humans (*Homo sapiens*) emerged, they were not alone. They were living in parallel with other hominid species such as Neanderthals (*Homo neanderthalensis*) and *Homo heidelbergensis*, which subsequently became extinct. (Whilst these *Homo* cousins had already been around for a few thousand years, they continued their existence for some time into the *Homo sapiens* era.) Interestingly, whilst shorter and stockier than modern humans, Neanderthals had brains that were just as large as, and sometimes larger than, our own, although they were more elongated in shape. Researchers have observed that whilst both modern humans and Neanderthals are born with relatively elongated braincases, only humans go on to develop globular-shaped skulls as they grow. The researchers hypothesise that the pattern of modern-human brain development after birth is quite different from that experienced by Neanderthals, accounting for the difference in skull shape changes during development, and suggest there must have been associated cognitive differences that distinguished modern humans.

Bigger, bigger, bigger . . .

Whilst our brains have tripled in size since the time of our earliest ancestors, the speed at which brain size increased has not been constant. The evolutionary changes that led to the emergence of modern humans were relatively slow in comparison to the speed at which modern humans themselves went on to evolve into the beautiful specimens we are today.

Early changes in the brain took millions of years to embed in our evolution, and it is not until about 800,000–200,000 years ago that the increase in brain size became rapid. This rapid change is thought to be associated with dramatic environmental shifts that were then occurring. Significant climate fluctuations, from very wet to very dry conditions, made the environment unpredictable. It is thought that having a larger, more evolved brain may have been an advantage to a species trying to negotiate dramatic environmental changes, having to adjust to periods of famine, then plenty, and back again. Perhaps the individuals who were able to survive such environmental challenges were those who were able to adapt: they could change their behaviour in order to survive under differing conditions. This suggests that their brains had the capacity for problem-solving, to change tack to find food or shelter in different ways, for example. It may also suggest that they had better memories; perhaps they were better able to remember and recognise alternative signs or locations of food or shelter in and beyond their local environment.

Maybe such individuals were better able to communicate with each other and work together to survive. Maybe their brains were more effective at driving automatic bodily functions, such as temperature regulation, and they were therefore just physically better able to survive. Of course, they were also able to reproduce and raise their offspring effectively to pass on any traits that gave them an advantage; that, after all, is how evolution works. There are many possible theories as to what advantageous traits gave the successful individuals the edge.

Whilst we don't yet know why the brains of early humans began to increase in size in the first place, we can see shifts in both physical form and ability as species evolved. All the time our brains were growing, our bodies were evolving too, as we grew taller and lost our body hair. Compared to our earlier ancestors and contemporaneous cousins, modern humans evolved a lighter, more slender skeleton and very large brains that are encased in a thin-walled skull. Our facial features are more delicate and we have smaller teeth. All these developments came about as our lives changed and our anatomy evolved to better meet our needs. There was a decreased need for physical strength for climbing and fighting as we became better adapted to living on the land, creating shelter and living more harmoniously with our neighbours (standing us in good stead for eventually organising community barbecues and coffee mornings). Jaws and teeth got smaller as we developed tools to do some of the jobs they used to, such as cutting meat, and we started to cook food, making it easier to chew. As our jaw took up a smaller proportion of our head, our skull may have evolved to expand into the space.

Eventually there came a time when our ancestors were no longer just focused on survival, but their abilities and interests were escalating. They became better able to exploit their environment for an improved life, leading to further survival advantages. Early humans learned to make and use tools, which in turn enabled them to do many other helpful things, and they discovered and harnessed the power of fire for keeping warm and cooking food. As their cognitive abilities evolved they became adept at living in groups, presumably being able to communicate effectively, negotiate and create social structures. Tools, clothing and other man-made objects became more sophisticated as both fine motor skills and thought processes advanced. Artefactual evidence shows that early *Homo sapiens* produced more advanced tools than other hominid species living at the same time, and he also created art, such as that seen in cave paintings. Further improvements, such as the domestication of animals, the introduction of farming, transport and trading and the building of complex societies, all stemmed from an increasingly sophisticated brain. As you can imagine, all of these developments in the history of mankind could only be achieved with concomitant improvements in the brain.

As you can see, the brain was not just getting larger; it was building new capabilities and investing in key areas that were proving the most useful. Whether new abilities evolved as a result of changes in the brain or whether the brain changed in response to the need for new abilities is as yet unknown. However, as our brain evolved, not only did it deliver upgrades in cognitive abilities, but more basic functions were also transforming. It is likely that as some areas of the brain became more important, they grew (over many generations) at the expense of the parts that were becoming less significant. Parts of the brain linked with aggression and other more primitive capabilities, for example, were no longer as prominent. Individuals who inherited smaller versions of brain parts that are needed for these functions may instead have had more brain room and energy to dedicate to more sophisticated cognitive and fine-motor abilities, perhaps giving them a survival and mating advantage. Those who could communicate and negotiate more effectively and produce tools, whether to forge objects, to hunt or to protect themselves, were likely to have had the edge in survival over their more primitive peers.

. . . pop! The bubble burst

All good things come to an end eventually. The human brain in its modern form emerged at least 200,000 years ago; the point at which it seems to have stopped increasing in size. Our brains finally reached a whopping 1,500 cm^3. You might think that all has been hunky-dory since then and our brain has been coasting along nicely. Not so. While we basked in around 190,000 years of physiological calm, little did we know that the tide was about to turn. Our brains started to shrink. As a species, we have lost around 100–150 cubic centimetres of brain since its peak. This is equivalent to the size of a tennis ball. This significant decline in our brain size relative to our body size has occurred over the past 10–15,000 years. But why? Is it because our brains have adapted to changes in lifestyle and they no longer need to be so big? Perhaps they are now more efficient at processing information – cast your mind back to how huge and basic mobile phones used to be and consider how comparatively slimline and powerful they are now.

John Hawks, a palaeoanthropologist at the University of Wisconsin-Madison, believes that the decrease in our brain size is indeed a sign that we are becoming more intelligent. As a large brain requires a lot of energy to function, a smaller, more streamlined one that is more efficient and powerful could use less energy but deliver greater cognitive ability. However, some people hypothesise that the brain would be more powerful if it were larger. Michel Hofman of the Netherlands Institute for Neuroscience argues that at a size of about 3,500 cm^3, corresponding to a brain volume two to three times that of modern man, the brain would reach its maximum processing capacity. He suggests that the larger the brain grows beyond this critical size, the less efficient it will become, thus limiting any improvement in cognitive power. So, according to this theory at least, the brain should have continued to grow.

In that case, what has happened to us? Perhaps we went into decline as our lives changed and we no longer used our brains for all the functions for which they were originally designed – a 'use it or lose it' effect. Some researchers do believe that the shrinkage is an indication of our cognitive decline as a species.

Researchers from the University of Missouri investigated how cranial size changed as *Homo sapiens* adapted to an increasingly complex social environment. They found that, around the globe, while population numbers were low the cranium kept getting bigger, but as populations became more dense it started to get smaller. They concluded that as complex societies emerged, people no longer had to be so intelligent to survive, so the brain started to decline in size. Previously, less intelligent (or worse adapted) individuals would have been more likely to die prematurely, or at least fail to find a mate and pass on their genetic legacy. In complex societies, however, it is thought that these individuals might be propped up by others and their genes passed on, thereby watering down the intelligence of the species as a whole. In order to ensure a continually developing intelligence for our species we would presumably have to be less nice to each other. Society would need to be structured in such a way that those who had less favourable traits to offer should be unable to survive long enough to produce offspring, or at least be unable to find a mate. A highly unpleasant and unacceptable prospect.

Chris Stringer, a palaeoanthropologist from London's Natural History Museum, suggests that some of the brain volume shrinkage relates to the fact that humans have got smaller overall in the last 10,000 years. It is possible that with the climate warming up we no longer needed to be so bulky in our form. In addition, he points to the fact that big brains use a lot of energy to maintain themselves, which isn't always necessary. He proposes that many of us are just fine now with smaller brains as we no longer need to store so much information; we have computers and gadgets for that, and we suppose that previous generations had books, songs and folklore which might have served a similar purpose. He describes this as a kind of domestication effect. Just as domesticated animals have smaller brains than their wild counterparts because they no longer need the cognitive abilities associated with hunting or protection, we humans may also have lost brain power as we have become increasingly domesticated.

In order to understand this brain shrinkage, we need to decide whether the shrinkage was wholesale or whether particular parts of the brain were withering away as they became less important. A group of Chinese scientists have found that whilst our brains have been shrinking over the last few thousand years, this shrinkage was not universal. Their evidence suggests that there is one part that has actually continued to increase in size: the frontal lobes. This part of the brain is involved in a host of functions, such as motor skills, problem-solving, judgement, language, memory, mood and emotions, social behaviour, and more.

So, is the brain still shrinking or is it continuing to change in form? Perhaps some parts are growing and some parts are shrinking? As Chris Stringer has neatly summarised things, it is plausible that the modern human brain is smarter in some ways and less intelligent in others, but overall it is more docile. Depending on your favoured theory, the brain has either already had its heyday or is increasing in complexity and specialisation.

Are we special?

Regardless of whether our brains are on the decline or in the ascendance, we are what we are and have to make the best of it. The average modern human adult brain weighs around 3 lb (1.36 kg). Whilst it accounts for approximately 2 per cent of total body weight, the human brain consumes a whopping 20 per cent of total body energy resources, due to the hungry neurons and their high metabolic requirements. But is our brain anything special? 'My cat seems pretty intelligent despite her small brain. So what exactly am I getting for all my extra brain?' asked a friend. Good question. Is all this fuss and using up so much energy to grow our extra brain really worth it?

Humans are just one of several species of great ape. Our closest living relatives are the chimpanzees, with nearly 99 per cent of our genetic code being the same as theirs. However, as we mentioned earlier, our brains are about three and a half times bigger than those of chimpanzees. Gorillas and orang-utans, whilst having bodies at least as large as humans, have brains that are only one-third the size of ours.

We share a number of features with our ape cousins. All primates have a similar arrangement of body parts, both internal and external; share the same bone structure; and all have forward-facing and closely spaced eyes, enabling excellent vision. Primates rely heavily on their sense of vision and in compensation have a relatively poor sense of smell in comparison with other mammals. We are all very good with our hands and feet and can manipulate objects skilfully, and typically have only small numbers of offspring. Yet despite all this, humans are quite different.

Although it remains a scientific challenge to explain the exact evolutionary changes leading to our unique behavioural abilities, there does seem to be some association between a big brain and increased intelligence. After all, a big brain seems pretty important for processing multifarious functions. One study investigating carnivorous zoo mammals found that brain size correlated positively with how successful a species was at problem-solving; i.e. the bigger the brain, the better at solving puzzles. However, it is unlikely that its size is the only distinguishing feature of the human brain. Elephants and whales, for example, have considerably larger brains than our own, yet haven't developed the ability to carry out intricate heart surgery or build a space rocket.

In the nineteenth century, scholars became fascinated by studying the brains of notable figures to see if they could account for their success from their having bigger brains. While it is reputed that both poet Lord Byron and Lord Protector of England Oliver Cromwell had particularly large brains, these early researchers found an enormous variation in brain size among eminent scientists and scholars, with many having brains of unremarkable dimensions. There is, therefore, a mismatch between the human species' exceptional cognitive abilities and its unexceptional brain size compared to much larger animals.

Human brain size must matter to some extent, otherwise why has it bothered to evolve so significantly over millions of years? However, overall size clearly does not entirely account for our species' abilities. So if it is not overall brain size that matters most, perhaps it is the way it is organised?

Is brain composition all a matter of matter?

Agatha Christie's Hercule Poirot famously credited his 'little grey cells' for his success in crime-solving. Presumably he was referring to the intellectual importance of the grey matter in his brain, but was his unswerving faith in it justified? Is it something about the grey matter that sets us apart from other species? After all, how many detective agencies do you know of that are headed up by a marmoset?

The brain is made up of two kinds of tissue: grey matter and white matter. These can be seen with the naked eye. Grey matter is densely populated by neurons and synapses and has long been thought to be critical for intelligence. It is largely confined to the thin outer surface of the brain known as the cortex. In addition, grey matter is comprised of other cells called glia, which provide both physical support and beneficial nutrients. White matter, on the other hand, is located deep within the brain and is made up of bundles of axons called tracts. The axons are 'myelinated', i.e. wrapped in fat, which acts as an insulator to help with transmitting electrical signals quickly across distant parts of the brain and from the brain to the spinal cord to connect the brain with the nerves in the rest of the body. It appears white because of the high fat content. In essence, grey matter processes information whilst white matter relays information to and from grey-matter areas, and is sometimes referred to as the brain's super-highway.

Recently, as better tools have been developed to study white matter in the living brain, researchers have started to realise that white matter health is very important for all kinds of intellectual function. The need for different parts of the brain to connect with each other and effectively pass on information is vital, and that's where white matter plays a critical role. Whilst the development of grey matter peaks in adolescence and then reduces over the subsequent years, white matter continues to develop into our twenties and possibly thirties, and it seems that it can also alter in structure in response to new learning experiences.

As mammal brains increase in size across different species, they gain proportionally more white than grey matter. In humans, white matter makes up about 35 per cent of total brain volume, which is the highest proportion among primates. (Interestingly, in pygmy marmosets this figure is only around 9 per cent, perhaps accounting for why they haven't got their detective agencies up and running yet.) Given our better understanding of the importance of white matter, maybe this explains why larger mammals are more intelligent. 'But that can't be right!' we hear you cry. 'Dogs/pigs/rats are highly intelligent and they aren't that big. And anyway, surely this links back to bigger being better, and we already discounted that as being the whole answer.' Indeed. There may well be more to it than simply the volume of white matter in the brain.

Human evolution may have involved an increase in both brain size and number of neurons, but only a moderate increase in body size in comparison with other primates, whose body size increased significantly as we diverged from a common ancestor. It is possible that whilst evolution in humans (and their earlier ancestors of the *Homo* genus) favoured increases in brain size, it was increases in body size that were promoted in other great apes. Brazilian neuroscientist Suzana Herculano-Houzel proposes that perhaps it is not metabolically possible to have both a very large brain and a very large body mass.

The relationship between brain size and body size is not as straightforward as one might think. As animals get larger, brains increase in size at a slower rate. Small animals actually have a large brain relative to their body size. Something called the encephalisation quotient (EQ) is sometimes applied as a way of illustrating how large a species' brain is in relation to what would be expected for an animal of its size (Figure 2 might be helpful for understanding this concept). The expected value is calculated from an average for similar animals. The bigger the value, the greater than expected brain size and, hypothetically, greater intelligence of a species. Using this approach, humans appear to be an outlier, with a brain that is up to seven times larger than expected for a mammal of their size and about three

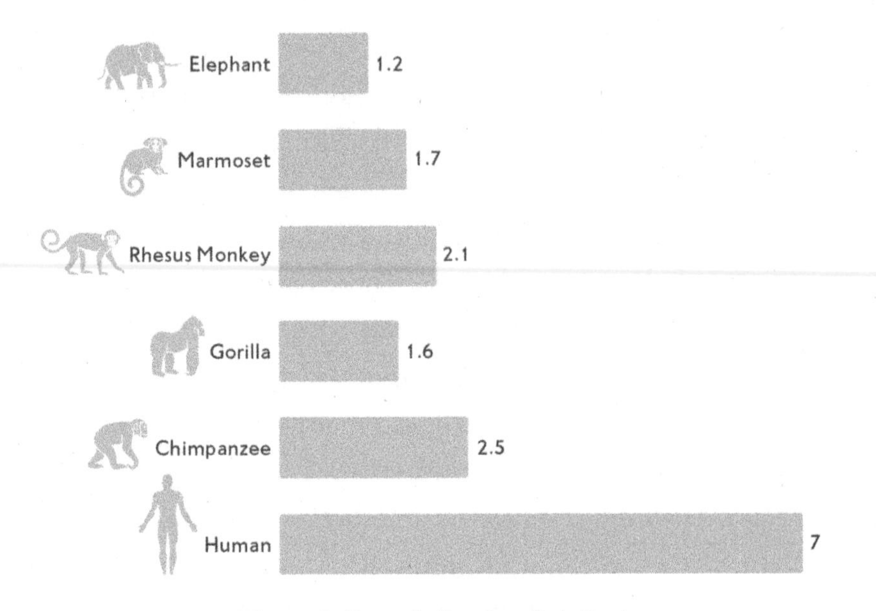

Figure 2: Encephalisation Quotient
Sources: Suzana Herculano-Houzel; Marino, L. Brain Behav Evol 1998; 51; 230–238

times larger than it should be for a primate species of their body mass. Even our early ancestors, with brains considerably smaller than ours are now, likely had higher EQs than modern-day chimpanzees. (Although it's worth pointing out that chimpanzees don't actually score that highly in comparison with other animals. Various species of dolphin, for example, score far higher. The relative intelligence of chimpanzees versus dolphins is a conversation for another day.)

Analyses of the EQs of a range of species have revealed that both herbivores and carnivores continually increased in brain size throughout their evolution, but at each stage of development the carnivores were always ahead. The EQs of predator species are typically higher than those of their prey. It's been suggested that carnivores require bigger brains to function and, as the brains of herbivores increased, the carnivores evolved yet bigger brains to maintain the differential. On top of this, according to palaeontologist and evolutionary biologist Stephen Jay Gould, primates have been ahead right from the start. Why this should be, however, remains up for debate.

In theory, just having more brain would allow a species to undertake more functions, and where there is more brain relative to body size, the 'excess' would enable individuals to undertake increasingly complex functions. However, EQ can't be the whole story either, since there are primates, such as capuchin monkeys, who score highly but are cognitively outperformed by lower-scoring species like the gorilla. EQ, like many other approaches, also has its flaws. It doesn't take into account factors such as the density and number of neurons, cortical thickness or the extent of brain-folding, all of which may have a bearing on intelligence. Interestingly, taking EQ alone as a measure of cognitive ability would have put Albert Einstein on a par with the dolphins, well below the average human EQ! He apparently had a smaller-than-average cranial capacity. However, when scientists took a close look at his brain they found that whilst the cortex was thinner than average, the density of neurons was greater. In other words, more neurons were crammed into a smaller space. So perhaps it is possible that the sheer number of neurons in our brain may have something to do with our intelligence as a species.

The human brain contains around 86 billion neurons. The cognitive advantage of the human species may simply lie in the total number of neurons its brain contains. According to Herculano-Houzel and colleagues, it is not brain size but the absolute number of neurons that imposes a metabolic constraint on brain-scaling in evolution: individuals with more neurons must be able to sustain the greater metabolic requirements to keep their brain operating effectively. It is thought that, whilst *Australopithecus* and *Paranthropus* species had a similar number of neurons to great apes (around 27–35 billion), there was a significant expansion in the *Homo* group, with 62 billion neurons by the time of *Homo erectus*. It has also been suggested that the significant increase in neurons estimated to have occurred between *Homo erectus* and the emergence of *Homo sapiens* was facilitated by the use of fire to cook food. This enabled greater calorific intake to feed the brain more quickly, freeing time for neurons to undertake more advanced activities.

However, when the number of neurons is considered in relation to the size of the brain, humans are not exceptional. We have the number of neurons that would be expected in comparison with other primates – we share a similar density of neurons to our primate cousins, but as we have bigger brains, we simply have more neurons. It might just be the total number of neurons that matters. Yes, we have the number that would be expected for our size, but this number is predicted to be the largest of any animal on Earth.

The brain is a powerful piece of hardware as a result of the way it has formed. Over time the cortex expanded enormously, but eventually would have been constrained by the skull, which was not expanding at the same rate. The solution to this was to fold in on itself and create more surface area so that it could continue to expand and develop in complexity. This is why our brains are wrinkly and look more like big squishy walnuts than big squishy hazelnuts. Research has found that the development of the cortex coordinates both folding and connectivity in such a way that brains are smaller and faster than would have been possible otherwise. Across different mammal species, the cortex becomes more folded as its size increases. In other words, the bigger the animal, the more convoluted the cortex. The human brain has the largest cerebral cortex relative to total brain size (somewhere between 75.5 per cent and 84 per cent), although other animals, such as the chimpanzee (73 per cent), horse (74.5 per cent) and short-finned whale (73.4 per cent), are not far behind. So, again, this feature alone is unlikely to fully explain humans' unique abilities.

There are many other avenues of research that are further unravelling the mysteries of human brain evolution at the micro level. It is likely that a combination of such factors, as well as those outlined here, is responsible for the uniqueness of humans. However, scientists still have a way to go in definitively determining exactly what it is about our brain that stands us apart from other species. Whatever the reasons, it is undeniable that humans possess exceptional abilities among the animal kingdom (more on this in Chapter 2), which must ultimately stem from differences in the brain.

Nobody's perfect

Our relatively enormous noggins have brought us a number of unique qualities over other species, including an innate smugness at being human. (Do any other species experience smugness?) But life isn't all peachy. There is always a downside, and big brain size is no exception. A number of disadvantages result from having such big brains.

First, and very importantly, a big brain means a big skull to accommodate it. Evolving to walk upright meant the human pelvis became narrower, to enable more efficient walking, which in turn meant less space for a baby's head. Many of us are only too keenly aware of just how big the human skull is, relative to our body size, when we have to push it out of a much smaller aperture during childbirth . . . (In fact, childbirth is more difficult for humans than for other species, and riskier too. Other primate mothers are able to reach down and assist their offspring by guiding them out of the birth canal and clearing mucus from their mouth and nose. Human mothers are unable to provide similar assistance due to the way their babies typically emerge from the birth canal.)

In order for humans to have such a large brain when fully mature, we necessarily give birth at an earlier stage in our young's development. Our babies arrive around six months early relative to when other mammals deliver theirs, with brains that are developed only to about 25 per cent of their adult size. In chimpanzees the brains of their young are about 50 per cent developed and in other primates this is nearer 75 per cent. Our children arrive into the world with a head full of unfused bone plates, which slide over one another to pass more easily through the birth canal and subsequently enable the brain to expand rapidly in the first couple of years or so before the bones eventually grow together. This is an ingenious way to enable the majority of brain development to take place outside of the womb, with fewer restrictions on growth and greater environmental stimulation, and offers an opportunity for cognitive development that other animals don't have. However, it does mean that our offspring are particularly vulnerable and helpless. Think of other mammals being born, such as foals and lambs, and how in a matter of minutes they get up on their feet and totter about. In comparison, our babies are unable to walk for many months, and are certainly unable to feed or look after themselves for many years. Being born so early in their development, human babies are far more dependent on parental support than other animals. In species-survival terms, all the time adults are looking after their offspring, they are consuming fuel but don't have time for finding food, building shelter, producing further offspring, etc. This is inefficient, and many other animals beat us hands-down in ensuring they produce and raise many healthy offspring to take forward their genetic legacy.

Another disadvantage of the human brain is that it consumes energy like it's going out of fashion. One-fifth of the calories we need to function are gobbled up by our brain, which is far higher than in many other animals. It is thought that it is all those neurons that use so much energy. The cost of this is that humans need to spend more time finding food, although it seems that learning to cook has enabled us to secure nutrients more quickly than other species and free up our time for other activities. However, research suggests that expanding our brains through evolution, and thereby allocating more energy to this part of the body, has come at a cost elsewhere. Some researchers believe that physical strength has reduced over time, with less energy being available for skeletal muscle. They speculate that the human brain and skeletal muscle co-evolved, continually balancing changes in energy requirements and availability.

Humans live a long time relative to other species, and are living longer and longer. Most of us now have easy access to energy-rich food and clean water, and we have medical advances to prevent us getting, or cure us of, many things that would have killed us in the past. As a result, our brains are subject to detrimental effects of ageing that other species typically don't live long enough to suffer from. Dementia, a syndrome in which there is deterioration in memory, thinking, behaviour and the ability to perform everyday activities, now affects millions of people worldwide. Although not considered a normal part of ageing, the biggest risk factor for developing dementia is age. Many domesticated pets now also seem to be developing dementia as they are living far longer with the assistance of their human companions, who provide them with food, shelter and medical care when necessary. In the wild, however, it is a different story. Wild animals don't live as long as their domesticated cousins – and even if they did, and subsequently developed dementia, they simply wouldn't survive. Living a long time isn't necessarily an advantage in the wild, as it is literally the survival of the fittest that counts; it's a young man's game, as they say.

We know a lot, but not that much

Humans have extraordinary abilities but the jury is still out on why this is. Whilst we don't yet know whether it is brain size or complexity that matters, it is probably a combination of both. It has taken millions of years to get our brains to where they are now, but was all that effort worth it? We know that the human brain has been shrinking, but we don't really know whether this means it is withering away or becoming more efficient, and what that might tell us about how much we really need. We seemed OK when the brain was bigger yet still seem OK now it is smaller; our lives, and corresponding skills, are just different. But are we getting the most out of what we have been given, or do we have untapped potential up there that we just don't know about? Now we know a bit about how our brains got to where they are today, we will go on to think about what the purpose of our current brain is and whether all of it really matters.

CHAPTER 2

Being Human: Why do we need our brains and what is it that really matters?

So we now know that the human brain has been on a pretty remarkable evolutionary journey and makes us rather distinguished among our fellow creatures. But it is now time to take stock and reflect upon our brain's main functions and what we really need from it.

We know that our brain is responsible for a host of vital functions, some of which can be specifically attributed to a particular area or nucleus. But not all of the brain has a specific function – or at least one that is yet known. Is this built-in redundancy, or an evolutionary hangover – do we really have more brain than we ever need or use? We now consider whether some parts of the human brain are like the appendix: seemingly unnecessary, troublesome and taking up precious space. Alternatively, what if we do need these bits of brain but don't yet recognise how each is vital to our very being? Since humans are more than just a list of basic physical functions, we look at what is in our brain that matters most: functions needed for basic survival versus those that seem at the core of what makes us human.

What's so great about the brain anyway?

In a nutshell, the brain is responsible for everything we do and who we are. For simplicity, however, we can think about what it does in some functional groups. First of all, there are those functions responsible for keeping us alive, such as controlling breathing, and those involved in movement, like balance and coordination. There are core functions involving how we sense and respond to stimuli, such as our moods, whether we are hungry, and temperature control. Then there are those functions responsible for making us who we are: from the way we learn and communicate, to our thoughts, judgement and social and creative skills.

Where do these many functions take place within the brain? This is not meant to be a textbook, and it would be pretty tedious to list every part of the brain in great detail. However, it is useful to have a sense of various parts responsible for major functions. So bear with us as we take a quick glance at the brain's big structures and main functions (returning to Figure 1 might help you to get your bearings). This won't take long, we promise!

One part of the brain that does not actually feature greatly in this book yet is utterly vital to our existence is the brain stem (the bit at the bottom of the brain leading into the spinal cord). It is responsible for many of the functions that are needed to stay alive: breathing, heartbeat and blood pressure. It is also involved in regulating our vision, hearing, sleep, feeding, facial expressions and movement.

The abilities to move properly and to maintain balance, posture and coordination are also imperative functions. It is the cerebellum (at the back of the brain) that takes charge of these, and its importance within the brain is becoming better understood – we will reveal more about this in Chapter 6.

How you and your body are feeling about things is largely dealt with by the limbic system (comprising a series of structures located in the middle of the brain, beneath the cerebrum). It contains glands that help relay emotions, and many hormonal responses are initiated here. The limbic system includes the amygdala, hippocampus, hypothalamus and thalamus. The amygdala is involved with the body's response to emotions, memories and fear, as well as recognition. The hippocampus converts temporary memories into permanent ones to be stored within the brain. It plays a major role in the storage of long-term memories based on knowledge and experience, as opposed to procedural memory, such as how to walk. It also helps people analyse and remember spatial relationships, allowing for accurate movements. The hypothalamus controls mood, thirst, hunger and temperature, and contains glands responsible for controlling hormonal processes throughout the body. The thalamus helps to control our attention span and keeps track of sensations, such as pain.

The cerebrum is the biggie (it's the big spongy bit that looks like the image of the brain we know and love). Not only does it take up the most space in the brain, it is also in charge of a huge list of functions, many of which we could consider to be central to what it means to be human. Needless to say, the cerebrum takes charge of a lot of things. From dealing with the five senses of vision, hearing, smell, taste and touch, and understanding and formulating speech and language, to processes involved in physical and sexual maturation, movement, and libido and hormones. In addition, the control of all higher-level functions is taken care of in the cerebrum. Problem-solving, abstract thought, creativity, reflection, judgement, initiative, inhibition, behaviour and some emotions are all under the control of the cerebrum. It is here that we sense fear, appreciate music and gain a sense of identity; this is where our personality stems from.

As the cerebrum is such a major part of who we are, and takes up a good chunk of discussion in this book, it deserves a little additional detail here. Within the cerebrum, certain parts are highly specialist, uniquely responsible for their primary task, whereas others are generalists, working in collaboration with other regions to serve a variety of functions. Amongst the specialists is the fusiform gyrus, a ridge on the underside of the cerebrum whose job it is to recognise faces. If it suffers damage, a condition called prosopagnosia, or 'face blindness', results, where the sufferer has intact vision but great difficulty in recognising familiar people. Another specialist is the area responsible for representing sensation: the so-called somatosensory cortex. We know a lot about the function of this region thanks to the pioneering neurosurgeon Wilder Penfield, who worked in Montreal in the 1930s, 40s and 50s. While conducting surgery on patients with intractable epilepsy, he saw an opportunity not just to help his patients but to map the functions of different parts of the brain. He conducted his operations under local anaesthesia so that the patients were awake and could talk to

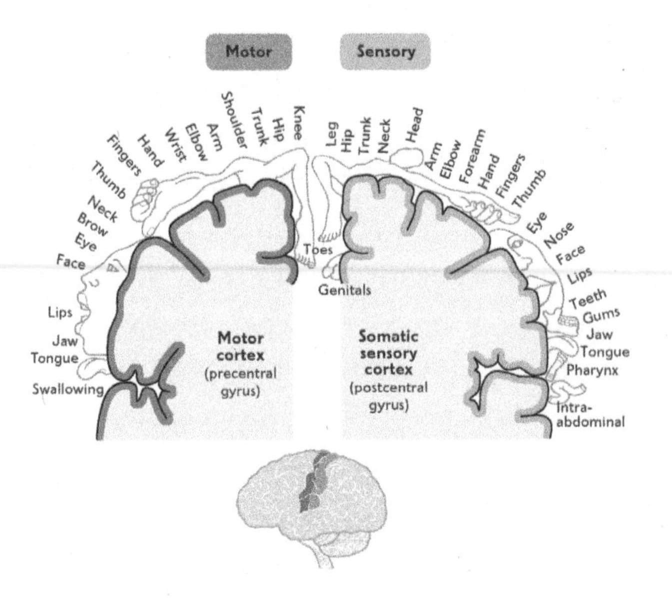

Figure 3: An example of a homunculus

him. He would electrically stimulate the surface of the brain and ask patients to describe their mental experience. So the operating theatre – or more specifically the brain of the surgical patient – become an opportunistic research laboratory. One of Penfield's key findings was that by stimulating the somatosensory cortex he could reliably generate sensations localised to specific bodily parts. He also discovered that the area of the brain surface devoted to processing sensations in body parts is not proportional to the area of the body surface, but rather how densely packed with nerves that region is. So the tongue and fingers, which have large numbers of nerves to allow for very refined sensations, take up a disproportionately large part of the brain surface. This is often represented by an image called a homunculus, which maps brain regions to a rather distorted-looking representation of a man with particularly large hands and lips, for example, in comparison to small legs and feet. You have probably seen something like it at some point (see Figure 3). Not all parts of the cerebrum play such a specialised role, and there are many parts that work more flexibly as a team.

Of course, the brain parts that we've highlighted above do more than the functions we've listed and there are many more parts that haven't even been mentioned. The brain is so much more than a mere catalogue of structures and functions; it is in fact a busy network of signals and activities that works as a whole to produce the complex entity that is the operational human being.

As you can see, the brain is extremely busy doing an awful lot of important stuff. However, there may be some parts that are just coasting along for the ride. If so, why are they in the brain at all?

Busy doing nothing?

Humans have been pondering the purpose of their brain for a long, long time. An Ancient Egyptian document, the Edwin Smith Surgical Papyrus, from the seventeenth century BC, is thought to contain the earliest reference to the brain. Fascinatingly, the papyrus gives the first accounts of several forms of brain injury and their associated complications. However, it seems that Ancient Egyptians didn't think much of the brain. As far as we know, they thought it principally passed wet mucus to the nose!

Despite thousands of years of study, we still don't know exactly what all of the brain does. Take the posterior cingulate cortex, located right in the middle of the brain. Some research suggests that it plays an important role in cognition, but no one agrees on what this role is. Ideas include a role in autobiographical memory or planning for the future, or maybe helping regulate our focus of attention. It is a highly connected and metabolically active brain region – suggesting that it's doing an important job – but we've yet to truly get to the bottom of exactly what this is.

The claustrum is another part of the brain whose function is enigmatic. It's like a small, thin sheet, with a volume comprising just one quarter of a per cent of the cerebral cortex. It may play a role in facilitating the wide travel of information to synchronise cognitive, sensory and motor messages, although there is little good evidence to support this theory.

And then there's the dorsal cochlear nucleus, which is located on the surface of the brain stem. As its name suggests, it seems to have something to do with the auditory pathway. Yet although it appears to be linked to tinnitus, what it actually does is still unclear.

Whilst the above represents just a few examples of brain structures whose function is yet to be determined, it is also worth pointing out that there is still much to learn about many other parts of the brain that already have identified functions. We need to know more about the basic biology and physiological mechanisms taking place in the brain, how different parts relate to each other, how they can cause ill-health or change as a result of it, how they can compensate for injuries elsewhere, and the extent to which we can help the brain to repair when there is a problem.

It is of course possible that there are parts of the brain that are evolutionary hangovers we no longer really need. There are examples in the human body of things that are no longer essential to our modern way of life, such as body hair and wisdom teeth, so why should the brain be any different? We've long thought that the appendix was one such body part that is a relic from a primitive age yet, funnily enough, it turns out that the appendix may still have a role after all.

For a long time, it was thought the appendix had originally been involved in the digestion of cellulose, found in plants, which our ancient ancestors consumed in significant quantities. We no longer eat the same diet, and so it was assumed that, without a pressing need for this organ, evolution had withered the human appendix. It just, literally, hangs around. More recently, however, researchers have come to believe that the appendix has greater significance than we've previously given it credit for. There is evidence to suggest that it protects our internal environment by helping remove waste matter from the digestive system, regulating pathogens, serving as a store for beneficial bacteria, and possibly producing early defences against disease. Likewise, just because there are areas of the brain that haven't had a function identified yet doesn't mean they don't have one. It may just be that we still have work to do in understanding what those functions are.

I am human; hear me roar!

But what about the parts of the brain we do know about? Which are the bits that are at the crux of what makes us human? Aside from those parts of the brain whose primary functions keep us alive, there are parts whose functions mark us out from other species and from each other; they are what makes each of us special.

The building blocks of brains are pretty much the same in all animals, but their numbers and how they are put together varies widely. For example, some animals have very large olfactory bulbs, associated with their exceptional sense of smell, whilst others, like humans, have large areas of the brain devoted to vision. The way human brains have developed means that we can do various things that other species cannot but, conversely, there are things that they can do that we have lost the ability for as our brains changed over time. Whilst there are many obvious physical differences between us and other members of the animal kingdom that point to different abilities – opposable thumb, bipedalism, lack of wings and gills – it is the differences in mental processes that interest us here.

Humans have developed highly sophisticated powers of communication. Our complex vocal cords and muscular tongue, among a number of anatomical differences, give humans extraordinary language abilities. In addition, mutations in the FOXP2 gene seem to have played a key role in the evolution of human language. This gene has been found to signal important activity in the brain during embryo development and it is thought that the human version of the gene contributes to our ability to learn to talk when young, by giving us particular control over our mouths.

Not only is our spoken language advanced; we have also developed written communication and sign language. In addition, we are able to communicate our own, and understand each other's, intentions, and can work cooperatively to achieve shared goals. While some insects, such as bees and ants, work together to a common purpose, in much of the animal kingdom cooperation between unrelated individuals is rare. Of course, humans don't always need to use language to make this happen. For example, one human can point to indicate an item they want and another human can understand this and help them by passing it to them. By this simple gesture of pointing, both the individuals share an understanding of the goal and know what is needed to achieve it. It is suggested that non-human species are at a disadvantage when it comes to cooperating effectively for no immediate gain to themselves, as they lack both cognitive abilities regarding future events or benefits, and language skills that enable cooperative working.

Our intelligence and cooperation as a species have enabled us to create complex societies and technologies that expand what we are able to accomplish. We can fly across the world in twenty-four hours, or carry out complex medical procedures with the assistance of sophisticated imaging techniques. It is arguable that, whilst other species often demonstrate extraordinary abilities, they have not evolved to an extent comparable with humans.

We use our abilities to think and contemplate, and our deductive powers and technology to look back in time and forecast into the future. We explore the depths of the ocean and the far reaches of outer space. We reflect on our place in the world rather than just using our energies to survive. In addition, many humans believe in a deity, or at least some form of spirituality, which again requires a brain capable of more than mere survival functions. Whilst we can't say for sure that no other species spends time reflecting on its own situation and what else may be out there, there is currently no evidence to suggest that they do. Although giant tortoises often look pretty contemplative . . .

Whilst it was long thought that empathy was a uniquely human trait, it has since been discovered that there are examples of other animals displaying empathetic behaviour. However, Robert Sapolsky, a neuroscientist at Stanford University, believes that where humans are unique is in the ability to feel empathy for things that are not even real. Just as we are able to consider our place in the world, we are also able to use abstract thoughts and even have emotions in relation to them. For example, we may be brought to tears watching the misfortune of a cartoon character on a screen. We are able to use metaphors and analogies, and have physical reactions to mental abstractions, such as feeling physically sick when confronted with something that is morally revolting.

We also present the abstract, and communicate, through our creativity. Our ability to both produce and appreciate art, music and story-telling is seemingly unparalleled in the animal kingdom and has been part of human behaviour for thousands of years. Many studies have demonstrated significant physical changes in our brains when exposed to the arts. For example, some research has found that music training in children can result in long-term enhancement of visual-spatial, verbal and mathematical performance. Studies suggest that humans' inordinate capacity for creativity likely reflects the unique neurological organisation of the human brain. Furthermore, since art is produced spontaneously only by humans and is ubiquitously present in human societies, any insight into artistic creativity can help us understand the neural underpinning of creativity in general.

Despite all these claims of human brilliance, researchers Gerhard Roth and Ursula Dicke from the Brain Research Institute at the University of Bremen believe that all aspects of human intelligence – with the exception of sophisticated language – are present at least in rudimentary form in non-human primates or some other animals. This is not a radical new contention. Charles Darwin believed that our intelligent behaviour developed from the primitive instincts of our non-human ancestors, and that the difference between human intelligence and animal intelligence was a matter of degree, not of kind. 'My object in this chapter is to shew [*sic*] that there is no fundamental difference between man and the higher mammals in their mental faculties,' he wrote in *The Descent of Man and Selection in Relation to Sex*.

There are of course many things humans are unable to do that other species excel at. We can neither fly nor live underwater, and the field of dentistry thrives on us being unable to replace our adult teeth, which sharks and reptiles can do. Whilst we have the intelligence to design increasingly effective prosthetics, we are yet to naturally regenerate a lost limb like a salamander can. Extreme climates would kill us off in droves but there are plenty of other organisms that cope just fine, whether in the Arctic tundra or searing heat and cold of the desert – think Arctic fox, walrus, vulture, scorpion. We can't see, hear or smell as well as many other animals. Whilst humans see things within the red-to-violet-light spectrum, some creatures, like bees and certain species of deer, can actually see beyond that spectrum and into the ultraviolet range. Humans don't have the ability to use echolocation or sonar, like bats and dolphins, and even pigeons can hear sounds at much lower frequencies than us. In many carnivores, the part of the brain that is devoted to smell is much larger than in humans. We bombard ourselves with public health messages about eating at least five portions of fruit and vegetables a day, in part due to the fact that our bodies cannot manufacture vitamin C. Other animals, such as cats and dogs, are able to do this, which is why you don't often see them begging for a segment of that orange you are eating. We don't run or swim anywhere near as fast as many other creatures and we don't use the Earth's magnetic field to migrate, like birds and turtles. The brain controls all such abilities and where each features prominently in an animal's life the brain will be composed in such a way to ensure there is adequate support to enable it to happen.

The same but different

We may differ enormously from other species, but it doesn't mean that we humans are all the same. Despite sharing most of our biological characteristics and DNA sequences, there is still great variation to be seen within the human species.

Of the two authors of this book, one loves mushrooms and one hates them. One loves cycling, the other one doesn't. However, both are female, of a similar age, grew up in a similar part of the world within similar cultures, and possess a human brain. So why do we differ at all? You only have to consider identical twins to appreciate that people with identical genetic make-up can develop differences in personality and behaviour. Studies have shown that the environment in which we grow and live, and our experiences, can shape who we are and how we behave; more on this shortly.

Many researchers are interested in how the brain facilitates individuality in humans. Some are looking at the variation in how our brain responds emotionally to everyday challenges. We know that, in any given situation, people may differ widely in their response to it: from calm and stoical, to taking bold action, to quietly whimpering, to running around and screaming hysterically. There is evidence to suggest that the brain circuits involved in our emotional responses are highly pliable and can alter with experience, subsequently affecting our temperament. Furthermore, because the brain is able to change in its response to events, psychological interventions can harness the brain's capacity for change, promoting positive behavioural changes that increase well-being and resilience.

There may be various physiological processes at work leading to individuality, from specific brain mechanisms, networks and molecular processes to genetic factors regulating the networks that control our behaviour.

We hear a lot about genetics determining who we are. However, there is also something called 'epigenetics' that is hugely influential. Let us try to explain it in brief. Proteins form the structure of our bodies and are essential in many of the processes that keep us alive. Genes are sections of DNA that provide the codes for specific proteins, and this is what we are referring to with the term 'genetics'. In other words, when we refer to a person's genetics we are really talking about the sequence of codes they've inherited that provide the instructions for who they are; for example, the series of instructions for brown hair, knobbly knees or colour blindness. 'Epigenetics' refers to how genes are read by cells and how or whether their instructions are carried out. Epigenetics is about external modifications to DNA that switch genes 'on' or 'off', and even the intensity with which instructions are carried out. These modifications do not change the genetic code itself but act as biological markers on top of the code, a bit like you might italicise or underline text in a document you're writing to emphasise an important part.

Epigenetics is involved in many normal bodily processes. All our cells have the same DNA and so, in theory, the same genes should code for the same processes throughout. But our heart cells need to do different things from our brain or gut cells, so epigenetics ensures that the relevant genes are switched on or off as appropriate so the various cells can get on with what they need to do. Unlike our genetic code, epigenetics can change and be influenced by the environment. Chemical pollutants, diet and stress can all result in epigenetic changes, for instance, and so, just as epigenetic markers are involved in normal human functioning, they are also linked to disease. For example, they may switch off a gene that usually protects against cancer. Epigenetic changes have been linked to a host of health conditions, including obesity, heart disease, various cancers and autism.

So, thinking again about the identical twins, we can now see one way in which two people can have the same genetic sequence yet be different in how they interact with the world. One study of eighty identical twins found that when they were very young their epigenetics were virtually indistinguishable, but as they got older, increasingly pronounced differences emerged. These differences were greater in twins who were older, had different lifestyles and spent more time apart, emphasising the importance of environmental factors in shaping their individuality.

If identical twins can experience significant differences, it's not hard to understand how the rest of us can differ hugely, as our life experiences on top of our genetics shape the core of who we are. In addition, as we discovered in the previous chapter, the majority of human brain development occurs after we are born, unlike in other species, allowing for significant environmental influence on how we develop, which may account for the huge diversity of individuality among humans, arguably not seen in any other species. In Part Two of this book we will delve further into normal variation within the human brain and tease apart fascinating differences between people and what they might mean.

We are all an enigma

Of course, we are all more than the sum of our parts. Although research has helped to identify which parts of the brain are associated with particular functions, there are many things that make us human that are not so easily explained. Features such as love and creativity must originate in the brain, but it is less clear how and why the brain generates these aspects of our being. Logically, it is understandable that the brain would develop or promote abilities that further human survival, such as better disease resistance, advanced problem-solving or more efficient energy use, but why it would enable music-playing, art appreciation or empathy towards strangers is not obvious. All such abilities take up brain space and energy resources and detract from survival functions, so what's the point of them? There may be many theories offering answers, but it would take another whole book to cover this in any depth. The point we're making here is that there is so much going on in our brains and so much we still don't know. And if we don't yet know what it all does and how so many brilliant things stem from it, the task of determining how much brain we really need clearly isn't going to be easy.

The next three parts of the book will present you with scientific evidence, extraordinary case studies and general musings to shine a light on what might be going on in the brain;

whether all of the brain is making a valuable contribution to human life; and where it is heading in future. And, of course, we will be constantly questioning whether we really need all of our brain anyway.

PART TWO

People Vary

What are the effects of normal variation in the human brain?

CHAPTER 3

Men: Does size really matter?

Just as they have longer arms and bigger feet, on average men have bigger brains than women: about one-tenth greater in size and weight. This is interesting because (again on average) bigger brains tend to be more robust: they are associated with better cognitive function and lower risk for some disorders like Alzheimer's disease. In Chapter 1 we discussed the various ways in which the size of the human brain has changed during evolution, and how increases in brain size are likely responsible for many of our finest human traits. So is it fair to say that the typically larger male brain is just . . . *better* than a woman's smaller one?

Before anyone gets upset, let's be clear: we don't think so. But there are a lot of consistent, and often subtle, differences between the sexes – not only in average brain size, but also in at least one cognitive function, many aspects of personality and behaviour, and most brain-related disorders, which we'll take you through in this chapter.

Sex-related differences in brain structure or function are interesting because they can tell us something useful about how genes, hormones, brain structure and social influences play out differently in male and female brains, from birth to death.

If you watch a certain kind of crime show you'll know that when you find a dead body, even if it's just an ancient skeleton, it's almost always possible to tell whether it belonged to a man or a woman. This is because the average man is not only larger than the average woman in many characteristic ways (height, weight, skull size), but also because many parts of the skeleton have distinctive differences in shape, such as the pelvis. Skeletons have become this way for good evolutionary reasons, often involving the differing roles of men and women in producing and then caring for offspring. Looking further up the body, is there any equivalent of that evolutionary signature detectable as differences between male and female brains? Here we explore whether how much brain you need depends, in part, on which sex you are.

Up close and personal

Imagine that you're a police pathologist and you've been given a brain that's – strangely – been found preserved but separate from its body. The lead detective wants to know if it's from a man or a woman. Would you be able to work that out? More than likely you couldn't based on size. According to a recent analysis of data from 15,000 people, the volume of male brains is on average 11 per cent greater than that of female brains. While this is a sizeable difference in average, there's a lot of variability within each sex, and so considerable overlap: there are plenty of men with small brains and women with big brains. Unlike in the skeleton, you wouldn't be able to work it out from gross differences in shape either. The large wrinkly upper part of the brain (the cerebrum) is 10 per cent bigger in men, along with a 9 per cent larger cerebellum (the small part of the brain towards the back and bottom of the skull) and 12 per cent more cerebrospinal fluid. But these differences in gross shape are too variable and too subtle to give you any confidence in guessing whether the brain belonged to a male or a female.

Picking up a scalpel, you cut the brain open and start to prepare some slices for examination under the microscope. Would we see any difference now between a male and female brain? If we looked at the tissue types, we might expect to see a higher proportion of grey matter than white matter in women (men have 9 per cent more grey matter but 13 per cent more white matter, so women have proportionally more grey matter than men). But the problem remains: our brains all differ, and the average differences between sexes are only a small part of that variability. You might be able to improve the accuracy of your guess by comparing the sizes of different structures in the brain to one another, because even when you adjust for differences in total brain size, some structures are larger, on average, in men than women, whilst others are proportionally larger in women. If you compared the sizes of a given structure or area on the left versus right side of the brain (the amount of lateralisation) you'd find some more clues, since in some areas there are differences in the extent of lateralisation between men and women. However, even with a good microscope, you'd be hard pushed to see many of the sex differences in a detached, pickled brain.

Things would be rather different if you were instead studying a living brain. There, using a combination of clever technologies, you could find sex-based differences in just about every aspect of how the brain functions at the cellular level. Take the hippocampus, for example. These seahorse-shaped structures are found deep in the temporal lobe, one on each side, and are very important in forming long-term memories, especially those involved in spatial navigation. If we zoomed in to the level where we could see individual neurons, we might spot differences between males and females in the dendrites, the branches that carry electrical impulses to and from the neuron cell bodies. If we watched the male and female hippocampal cells in action, we'd see differences in their functional properties too: the female ones would be more sensitive to some types of neurotransmitter and less sensitive to others. And the cells in male and female hippocampi would differ slightly in their response to external stimuli: from how much excitation a cell needs before it will fire, to how likely a cell is to be damaged if the brain's owner is exposed to long-term stress.

These sex-based differences at the cellular level are the basic building blocks that eventually lead to differences in behaviour. As we move from studying what cells do to studying what entire organisms – people – do, we find that behaviour starts to involve not just innate differences between males and females but also learned ones. Picking those things apart is no easy thing to do. But we like a challenge, so let's have a go.

Maths and other headaches

Women can't read maps or park cars. Men are bad at listening and talking about their feelings. Do such clichés about the different abilities of men and women tell us anything useful about male and female brains? Because whether the participant is male or female is recorded in probably every human research study that occurs, there are plenty of datasets out there that could provide evidence as to whether such sex differences exist.

Despite this, level-headed discussion of sex differences and their causes is not as common as you might think. For one thing, the only studies that make the headlines tend to be the ones that find dramatic sex differences, not the many methodologically stronger, larger studies that find no difference. For another, discussing the causes of sex differences is a political hot potato. Just ask Larry Summers, the eminent economist and then President of Harvard University, who in 2005 publicly commented that the lack of women at the highest levels in science and engineering could be due, in part, to a 'different availability of aptitude at the high end'. Professor Summers was forced to resign soon afterwards. However, he raised an interesting and entirely testable hypothesis. Are there really any basic differences in aptitude between men and women?

Summers may have been referring to the fact that, historically, male students have done better than females in exams in maths and similar subjects. Indeed, they still do so in some countries, especially countries where overall ratings of gender equality in society are low. However, the most recent and largest studies from the USA have found that there is no longer any meaningful difference in average maths scores between male and female students. Yet, even if the average scores are the same, there could still be differences in scores at the top and bottom ends of the distribution: men and women could have the same average maths scores if there were both more outstandingly good and more outstandingly poor results among men, for example. This 'male variability' theory has been around for more than a century: it was originally suggested in 1894 by Henry Havelock

Ellis, who noted that there were both more male geniuses than female ones, and more learning-disabled men than women.

Over the past couple of decades many studies have looked at whether young men score more variably on maths exams than young women, and several sets of researchers have independently attempted to synthesise the data. Overall, they conclude that there is slightly more variability in maths performance among male students than female ones, but that the difference is so small that it cannot possibly explain the very large sex differences seen in course and career choices followed in maths-heavy fields.

Even if the male variability theory were unquestionably true, Summers' comments would probably still cause offence because of their implication that there are inherent differences between men and woman in the uppermost level of maths performance. By using the word 'aptitude' rather than, say, 'achievement', he seems to be indicating that the difference is determined by biology, not society. As with most nature–nurture debates, it's hard to precisely pin down the relative contribution of the two. Since we cannot give a maths test to newborn babies, all tests of maths skills end up inherently testing not only an individual's aptitude in acquiring these skills, but also what they have been taught, how well they have been taught and how much they have been motivated to learn.

What is clear is that over the past fifty years female students in the US have made significant gains and have now essentially caught up with their male counterparts. It's extremely unlikely that biological determinants like genes or hormones have changed much over this time-frame. So the most likely conclusion is that the large sex difference that used to exist was driven by societal influences and attitudes, which can change, and have changed, very rapidly over this period.

Perhaps females were historically less often encouraged to take more advanced maths classes; perhaps the mostly male science and engineering faculties were biased towards young men in their recruitment and promotion practices; perhaps young women with high maths ability chose (or were diverted into) career paths that were more conducive to raising children. These are all much more plausible causes for any sex-related achievement gap – and, incidentally, issues that you would want a university president to be taking seriously.

Summers should have known that a better way to understand whether there are sex differences in 'aptitude' is to study performance on tests where societal influences, such as schooling, can have minimal impact. In other words, instead of comparing exam scores between male and female students, we should study more basic cognitive abilities. And we should study them as early in life as possible, before cultural and societal influences have had much opportunity to take hold.

The real test of an innate sex difference in aptitude would be that it exists universally across cultures, even among young children, and/or regardless of educational experience. There is probably only one cognitive ability that meets these criteria and, interestingly, whether or not Larry Summers knew it, it may well influence ability in some engineering-related skills.

So can women really read maps?

The one cognitive domain in which males seem to have a reliable advantage is in the spatial skill of mental rotation. The standard test for this is to show a series of drawings of three-dimensional objects, each of which must be matched with another drawing

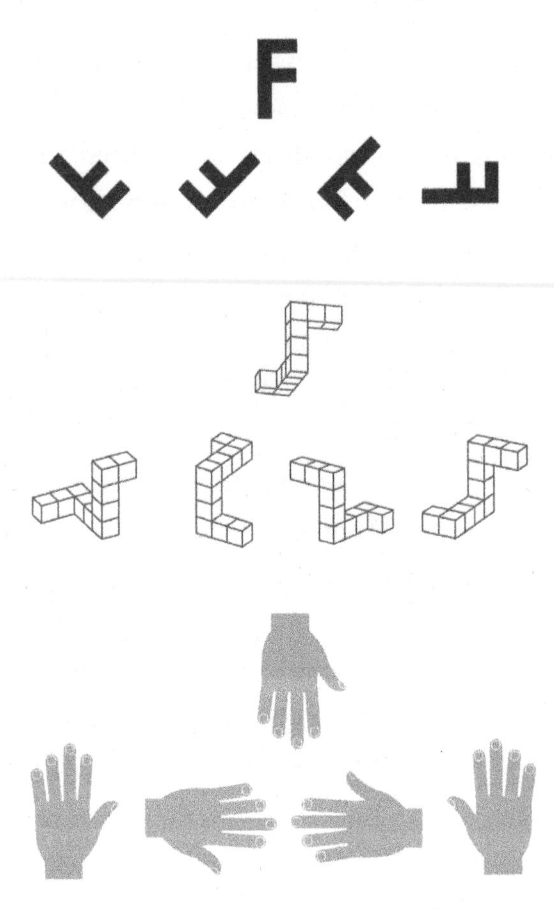

Figure 4: Mental rotation tasks

of exactly the same object shown from a different viewpoint, selected from other drawings that are nearly but not exactly the same, for example, drawings that show a mirror image of the object (Figure 4 provides some examples).

Across a large number of studies, males reliably outperform females on this test. To understand exactly how much by, we need some way to standardise the difference in scores of males and females across studies. To do this is a bit technical so feel free to skip ahead: the short answer is there's a 'medium to large' difference, by scientists' standards.

If you really want to know, one way we can express the difference across many studies between males and females is to say what variation in scores was observed in each study across the whole population, and then express the male and female difference in relation to that variance. One study into mental rotation is carried out in a workplace dominated by twenty-something-year-old software developers. Another involves stopping every third person in the street and asking them to take part. You'd expect the scores of the first study to be more tightly clustered, because they're a more homogeneous population, whereas the study on the street would include a much wider range of age and ability. But that's OK, we can still use both to tell us about male–female differences, as long as we can express the sex difference in both cases in terms of how big the difference was in each study compared with the spread of scores that was seen in that sample. We express this difference, standardised across studies, as 'effect size'.

Combined across many different studies of sex differences in mental rotation, the mean difference in scores is reported to have an effect size of around 0.6–0.7. What does that mean? Well, in scientific terms, 0.5 is defined as a 'medium' effect size and 0.8 as 'large'. Jacob Cohen, the American statistician and psychologist who first worked these things out, said that an effect size of 0.5 was large enough to be visible to the naked eye: for example, it's the difference in height between fourteen-year-old and eighteen-year-old girls. A score of 0.8 is more noticeable still: it's the difference between thirteen- and eighteen-year-old girls. For those who prefer numbers, an effect size of 0.65 would mean that a man chosen at random would be 68 per cent likely to do better on the mental rotation task than a woman chosen at random, but there would be a 75 per cent overlap between male and female scores.

This is quite a substantial difference in performance. Is it caused by sex-based differences in the brain? Well, one piece of evidence in favour of that is that several studies have now shown that male babies of around three to five months old seem to be able to do some 3D mental rotation of visual objects, which female babies of the same age largely cannot. Testing anything in three-month-olds other than their awesome ability to eat, sleep and defecate is of course pretty tricky, so all cognitive testing in this age tends to take advantage of one psychology trick. This is that babies look longer at things that are less familiar to them. So in this case, the testing consists of getting the baby used to a 3D object (or a picture or film of it) until the baby is familiar with it – bored of looking at it, you might say. The object is later presented again in a rotated position, and if the baby looks at it for less time than a completely new object, you can assume it's because the baby recognises it as the same object, despite its rotation. These are exciting but finicky studies to do, so they tend to be relatively small and hard to interpret.

Whether or not there are early aptitude differences between baby boys and girls, there's an alternative explanation that is also rather interesting. This is that boys practise mental rotation skills more frequently than girls, for example by spending more time playing sports and video games. These are thought to train relevant spatial skills, unlike mainstream school subjects, which do little in this realm. In one study, students were asked to play either an action-based video game or a puzzle game: after just ten hours of play, students playing the action-based game had significantly improved their mental rotation scores, and the women in this group had caught up with the men in the puzzle group. If such little training can allow women to 'catch up' with men, it's hard to conclude that sex differences in performance can be caused by much innate difference in ability, and more likely that they reflect differences in how boys and girls in the modern world prefer to spend their free time.

Personality and behaviour: what can you blame on your 'male' or 'female' brain?

As we grow up, early differences in temperament develop into lifelong traits of personality. Psychologists who study individual differences between humans usually agree that there are five major domains of personality that can be seen across all human cultures (and in some other species too). Of these Big Five, there are relatively consistent findings of sex differences in at least two: neuroticism, or the tendency to experience negative emotions; and agreeableness, the tendency to get along with others. In both traits, women score higher than men. Women also tend to score more highly on extraversion and conscientiousness, but these differences are less consistent. (The fifth factor, 'openness to experience', is not usually strongly differentiated by gender.)

These are very high-level descriptors of personality, however, and within them there are sub-factors that are deemed to be more stereotypically male or female. For example, there is considerable evidence that females show greater empathy and tender-mindedness, both aspects of 'agreeableness', and both conceivably linked to the characteristics needed to successfully nurture offspring. In contrast, males show more aggressive traits and more physically aggressive behaviour – and this starts early as toddlers begin to play with one another. When angry, men are more likely to hit things; women more likely to cry.

By now you're probably thinking: But is this really telling us anything about the brain? Since the vast majority of psychology studies are carried out among affluent Western cultures, couldn't this all just be an effect of the difference in social roles expected of men and women in these societies? Or instead you might be thinking, Well, that makes sense: males who are more aggressive will, throughout evolution, have been more likely to win the competition for mates, or perhaps just been more likely to have survived. To put it more broadly: an evolutionary perspective might argue that it has been evolutionarily beneficial (a.k.a. 'adaptive') for men and women to behave differently, in terms of either sexual selection or parental investment. A social-learning theory would postulate that men's and women's behaviour is shaped over a lifetime by rewards, punishments and role models.

There is certainly something to be said for both the social and evolutionary arguments, and indeed both nurture and nature are drivers of gender differences in personality and behaviour. What's of interest to us is, do these cultural and/or evolutionary influences show up in the brain? And if so, what can this tell us about what is and isn't necessary for a brain to work effectively?

Sadly, this is not an area that gets a lot of detailed neuroscientific investigation: historically, few people have been interested in funding expensive neuroimaging studies into something like how the brains of moderately agreeable people differ from those of moderately disagreeable ones. So we know relatively little about how much normal variation in personality can be linked with any differences in the structure or function of the brain. We know much more, however, about the brain characteristics of people at the extremes of personality or behaviour. Many psychological and behavioural disorders can in fact be viewed in that way: as an extreme variant of a behaviour that, at a less severe intensity, is neutral, or even desirable. For example, high scores on the personality trait of neuroticism predict an increased likelihood of going on to develop depression; high levels of impulsivity are associated with many forms of addiction and attention deficit hyperactivity disorder (ADHD); and obsessive-compulsive disorder (OCD) has been linked with very high levels of conscientiousness. So if we're looking for brain-based explanations for sex differences in personality and behaviour, one way in is to look at what is known about the role of sex in developing these disorders.

Who's more at risk?

Disorders that originate in the brain often occur more in one sex than the other. The links we noted above between normal and extreme variations in personality and behaviour sometimes give us a clue as to many of these sex disparities. Thinking about what we know already about neuroticism, you won't be surprised to hear that mood disorders such as depression, and anxiety disorders including panic and post-traumatic stress disorders, are all more commonly diagnosed in women. Thinking about aggression and impulsivity, you similarly may not be surprised that rates of use and abuse of illegal substances are higher in men, and that men are around three times as likely as women to experience a head injury. For other conditions the pattern is more complex, or it's less easy to imagine a societal-level explanation. Disorders that are thought to be caused by deviations in early brain development are usually more common in males: boys are four times more likely to be diagnosed with autism, and twice as likely to be diagnosed with a learning disability or ADHD. In schizophrenia and OCD, whilst the risk is broadly equal for men and women across a lifetime, males tend to be affected at a younger age. Interestingly, among later-life and degenerative brain disorders, there is no clear sex-related pattern; for example, more women develop Alzheimer's disease, but more men Parkinson's disease.

The reasons for these sex differences are varied. For some we simply don't know the answer; but for others there is a relatively simple explanation, which doesn't necessarily tell us much about the sex differences in the brain. One reason that Alzheimer's disease affects more women than men, for example, is because by far the biggest risk factor for developing Alzheimer's is getting old, and women tend to live longer than men. To understand what any of these sex differences tell us about the brain, we need to make sure they really do reflect differences in something about how the brain works, and not just about how other aspects of physical health, or social roles, differ between men and women. We might ask, for example, whether doctors are equally likely to give men and women a particular diagnosis if they are exhibiting the same symptoms, or are trying to describe the same internal experience. There may well be conscious or unconscious biases in the way we experience and talk about symptoms, and in the way doctors label our symptoms in the light of their training and personal experience. In reality, men and women experience life with differences in both biological and social influences, and for many disorders the two may work together in producing specific symptoms, or triggering illness at a particular time.

Consider the following two (fictional) case histories:

1. As a child, Susan was kind, sensitive and eager to please. She found the transition to senior school socially difficult, as many of her younger friends were left behind and her classmates teased her for being physically immature. This changed as she hit puberty, but she now felt self-conscious about her rapidly developing body. She suffered from heavy and painful periods, which often caused her to miss a day or two of school, until her GP prescribed her the contraceptive pill when she was fifteen. As a young woman, Susan suffered from episodes of low mood and anxiety but did well in school and went on to a good university, where she met her future husband. They married on graduation and she had her first child at the age of twenty-three. She suffered a period of postnatal depression, which was recognised early and successfully treated with antidepressant medication. Two years later she had a late-stage miscarriage and went back on to antidepressants. A year later, in discussion with her GP, she decided to stop taking them while trying to become pregnant again.

2. Susan's brother Ben was notorious even as a child for being a risk-taker. Born prematurely, he was always rather small for his age. At school, Ben was popular with other children, and had a reputation as the class clown. As he got older, his disruptive behaviour got him into trouble on a number of occasions, and this led to arguments at home. Leaving school at sixteen with just two GCSEs, Ben worked in a number of shops before starting at a local garage as an apprentice mechanic. Still living at home in his mid-twenties, Ben drank heavily at weekends in the local pub with his large social circle. He was briefly banned from driving after he failed a breathalyser test when caught speeding by police on a Sunday morning.

We don't know if Susan would have experienced the same symptoms of anxiety and depression if she had been born male. It does seem likely that some of her symptoms were caused or exacerbated by hormonal changes: we know that some women experience strong changes in mood related to their reproductive cycle, and that events like miscarriages, pregnancies and changes in the use of hormonal contraceptives can all trigger depression. We also know that poorer school performance and alcohol abuse are both more common in men who were born prematurely. And, like the rest of us, Susan's and Ben's mental health during young adulthood in part reflects the unique circumstances and events that arose in their lives, some proportion of which would have been different had their sexes been different.

Alcoholism and depression tend to run in the same families, though this may reflect common environmental risks rather than shared genetic ones. But genes are the most basic way in which males and females differ – so what role do they play in the sex differences in risk for brain disorders? It's a complicated question. All psychological and neurological conditions have some genetic basis, but relatively few of the genes that are known to increase risk for any of these disorders are actually located on the sex chromosomes. When you think about it, this makes sense: since only males carry the Y chromosome, no genes that are really important for brain function could be located there.

Let's take a closer look at a disorder with both a strong genetic basis and a large difference in prevalence between the sexes: autism. The genetics of autism is a subject of considerable research interest and, so far, hundreds of locations on the genome have been shown to increase risk, only a handful of which are on the X or Y chromosomes. Why so many genes? Well, perhaps in part because the sheer complexity of the brain means that an awful lot of genes are needed to control its development, structure and function. Moreover, if only one gene controlled one aspect of brain development, that would make it very vulnerable to problems: if a single mutation occurred in that gene it could be fatal if it affected one of the vital functions the brain controls, like breathing. So on the whole there tends to be some overlap and redundancy in genetic influences on the brain. What that means is that there are also a lot of ways that differing changes in the genome can give rise to similar changes in the structures or pathways in the brain. Whenever those changes in the brain cause a cluster of symptoms, including problems with communication and social interaction, we label that as autism.

Even if the majority of genes that increase risk for autism aren't found on the X or the Y chromosome, genes may still be very important in explaining the sex difference. For one thing, genetic variants located on the sex chromosomes may interact with those elsewhere on the genome to increase overall genetic risk in males. Or it might be that something about having two X chromosomes is protective: for example, differences in the way that genes on one X chromosome are expressed while those on the other are silenced might help protect females from harmful brain effects.

One recent large study suggests that there are considerable differences in the extent to which harmful genetic mutations affect males and females with autism. The researchers, led by Sebastien Jacquemont and Evan Eichler, looked at the DNA of nearly 800 families where one or more of the children had autism. By comparing the DNA of the affected child with that of the two parents, they were able to see how frequently a new mutation had taken place in the child's genome that did not match either parent's DNA. Such mutations are common, they are mostly harmless, and only a minority might be expected to have any effect in the brain. They also counted the number of genetic variants in each person's DNA that are known to be harmful in some way (for example by prematurely shortening the genetic sequence that defines a protein). When they compared males and females with autism the researchers found something rather surprising: females with autism not only carry more new mutations in their genome than males, but also three times as many harmful variants. This adds to the 'female protective hypothesis' because it implies that females have to carry a much more harmful genetic load before it affects them sufficiently for them to be diagnosed with autism.

A somewhat different explanation for the increased tendency of males to have autism has been put forward by the Cambridge University researcher Simon Baron-Cohen. He argues that there are differences between males and females in their tendency to empathise versus their tendency to systematise the world: females tend to show more empathy and are over-represented in caring professions such as nursing; males tend to have a stronger drive to create and analyse systems and are over-represented in professions like software engineering. In this model, autistic traits are seen in people who are both high in the systematising trait and low in empathy, and autism is seen as the 'extreme male' brain.

This theory is not universally accepted but it raises interesting questions. How would these differences arise in the developing brain? One possibility is that they are linked to the amount of testosterone present in the foetus, which can be measured by analysis of the amniotic fluid that surrounds the foetus in the womb. Differences in foetal testosterone are thought to drive many of the ways that developing foetuses differentiate according to their sex, and studies have shown that children with more autistic-like behaviours had higher testosterone levels in their amniotic fluid.

Another possibility is that sex differences in the rate and pattern of normal cortical development throughout childhood and adolescence may be disrupted in autism. This idea is interesting because we know that other developmental disorders seem to involve exaggerations of normal brain development. For example, children with schizophrenia, a disorder that shares considerable genetic risk with autism, show an acceleration of the normal pattern of loss of grey matter over time, alongside slower growth of white matter, than their healthy peers.

Let's just say 'it's complicated'

In this chapter we've run a quick survey of how and where sex differences seem to play out in the brain functions that matter most to our argument about what it takes to be a successful human. What have we learned?

When we looked at the fundamentals of how the brain worked, we noted that subtle sex differences can be seen all over the place. It is interesting to consider how much of the psychology and medical literature is based on research conducted almost exclusively on males. There is a simple reason for this: males (particularly young, white college students) have historically been the easiest people to get access to if you're a psychology or medical professor. But the bias is, even in modern research, surprisingly large: for example, fifteen males are included for every one female in studies that look at brain activity in autism using functional neuroimaging. In fact, sex differences are so common and potentially so important that the US National Institute of Health recommends that sex should be considered as a variable in all clinical studies. That means, for example, if you are developing a new drug for depression or ADHD or Alzheimer's, you should make sure you test it in both males and females – and check whether there are differences. Since this rarely takes place, we can conclude one thing: that we know much less about how the female brain works than we do about the male brain.

We have come across many examples where culture is a more likely explanation for sex differences than anything that's going on inside our brains. But of course we know that brains do differ in many small ways, and that genes, environments and, in particular, hormones drive sexual differentiation right from our earliest foetal days. Do any of these affect how much brain we need? We would argue not: these differences are all in averages and we see large, often very large, overlaps between the male and female brain everywhere we look.

Perhaps surprisingly, the most basic unit of sex difference, the X and Y chromosomes, hasn't shown up as playing a huge role in determining how brains function. Should we be surprised about that? It's worth remembering that whilst humans have twenty-three pairs of chromosomes, only one pair of these differs between sexes. Therefore, the evolutionary pressures which have selected for genetic variation among the other twenty-two might be expected to drive more similarities than differences between the sexes.

It's also worth noting that evolution might work on the brain in two ways: to cause differences between the sexes, or to compensate for them. The big drivers of evolutionary success, which are often summarised as sexual selection and parenting success, have clearly driven some exacerbation in the differences between men and women, for example in aggression and nurturing tendencies. But at other times, natural selection might favour variants in the brain that help make up for the substantial physical differences between the sexes. If you are faced with a stranger who is larger and stronger than you, charming him is more likely to pay off than assaulting him. As Marilyn Monroe is thought to have said, 'Give a girl the right shoes, and she can conquer the world.'

A VIEW FROM

Dr Lauren Weiss, Associate Professor in the Department of Psychiatry at the University of California, San Francisco, USA

Though she sits in a department famous for its psychiatric research, Lauren Weiss is a geneticist at heart. Her interests lie not so much in cataloguing the many ways that human behaviour differs, but rather in understanding the many complicated and surprising ways that the human genome does. In particular, she is interested in understanding what the genetic basis of autism and other developmental disorders can tell us about the sex differences seen in these conditions.

To do this, Laurie heroically leads her lab in two distinct directions. In one they look at what can be learned from statistical analysis of datasets containing DNA from tens of thousands of patients with autism. In the other, they study what can be learned under the microscope from painstaking laboratory experiments.

Laurie, you've been trying to work out what the sex difference that we see in rates of autism means for how autism develops. But first of all, are we sure that more boys are really affected than girls?

'That's a good question. It's definitely possible that there are biases in the way in which boys versus girls with autism are detected and diagnosed. What we do know is that all of the ways that we can measure autistic tendencies show a consistent and substantial sex difference, with more boys affected than girls. But there's no lab test or biomarker for autism; it's defined only in terms of behaviour. So we don't know if there are some girls out there who have the biology of autism but don't get diagnosed, and also don't get picked up by any of the behavioural scales we use.

'We also just know less about females with autism: any time we are recruiting participants or collecting samples for a study, we get maybe four times as many males as females because so many more males are diagnosed. And because how much we can learn depends on how big a sample we have, that means we always learn more about the biological mechanisms and genetic risk in boys than in girls.'

So what do we know about why autism is more common in boys than girls?

'The bottom line is we still don't really know! We do have some intriguing clues though.

'One line of evidence is about the general mechanisms by which genes control development, and how that differs between the sexes. In one recent study we looked specifically at regions of the genome that show a sex difference in the extent to which they are associated with body measurements that show big differences between males and females: things like height, weight, hip and waist measurements. These regions are not on the sex chromosomes, but they affect males and females differently: typically they have no effect in one sex but a big effect in the other. We took this list of genetic variants and looked at whether they seem to affect risk for autism – which is completely unrelated to these physical traits – and they do. So this is one line of evidence suggesting that the sex difference in autism is real, and also that it's related to some very general ways in which development differs between the sexes.

'A second thing we have realised is that females with autism seem to carry a greater load of big, bad genetic changes than males do. This doesn't seem to apply to the common, small differences in genetic code, but only to changes such as repeats or deletions of large sections of the genome. Females with autism seem to have more of these and they also seem to have a more severe form of autism, for example they tend to have lower IQ. This could indicate that females are generally less susceptible to autism, so it requires a worse genetic mutation (or other risk factor) to get it at all – and when they do it tends to be a severe form.

'Another possibility is that the same genetic risk might show up in different ways, as different symptoms, in males and females. So two people might have the same genetic variation, and that might result in autism or ADHD in a male, but an eating disorder or anxiety in a female. Now that we are beginning to get a better list of risk genes for most developmental and psychiatric disorders, it's clear that there is considerable overlap and that many of these genes are associated with risk for more than one disorder.'

Are there any examples where it's clear that genetic or environmental risk factors affect girls and boys differently?

'One good example of this are the RASopathies. This is a rare genetic set of conditions, each of which is caused by a particular mutation in a gene in the Ras-MAPK cell-signalling pathway. We know that a high proportion of people with RASopathies have autistic traits or autism: somewhere between 10 and 50 per cent, depending on which particular genetic condition it is. But unlike autism, if you have one of these specific mutations you will definitely develop the RASopathy, so it's a simpler genetic model we can use to investigate things like sex differences.

'The RASopathies occur equally in males and females. But what we recently found out is that, for some of these genetic conditions, the autistic traits are much more prominent in males, whilst in others they occur equally across the sexes. So this is a situation where known genetic mutations are having sex-specific effects in producing autistic behaviour.'

That sounds very promising. What's the next step for this line of enquiry?

'In our lab we've been working to develop a model of RASopathies that will allow us to test hypotheses about the effects of different risk factors in a petri dish. To do this we take skin cells from people with RASopathies and then, using a special technique, we can change those skin cells back into stem cells, and then induce them to turn into neurons. This allows us to generate an infinite number of neurons all derived from that one person's genome. Right now we're just working out how to measure things like differences in how the neurons grow, how they express different genes, and their signalling properties. The next step will be to start to look at how they react to different environmental or genetic influences that we think might be of interest in autism.'

How much brain do we really need – do we need it all?

'I don't know about the brain, but it's certainly impressive how much genome we can live without and still be functional. Sometimes when we're looking in datasets from healthy volunteers, we see really large deletions in the genome: if we'd seen them in a patient sample we'd have said that is definitely the cause of the problem. But in these people it's having no obvious effect, presumably because other aspects of their genetic background are protecting or compensating for its effects. We know in some genetic disorders people can lose a stretch of genome with maybe fifty genes on it and still be walking and talking and doing pretty well.'

CHAPTER 4

The Prime of Life: When exactly is that?

In 1905, Albert Einstein published four papers that changed our understanding of physics for ever. In this one miracle year, he described mass–energy equivalence with the equation $E=mc^2$; outlined the basis of wave-particle duality; defined Brownian motion; and proposed his theory of special relativity. He was twenty-six years old at the time, an age at which cognitive processes are fully developed but when the brain has yet to experience much deterioration. Was Einstein in his neurological and cognitive prime?

Theoretical physics, like maths, is a field in which the greatest talents are often said to have done their best work before they are thirty. There are certainly many examples of these early prodigies. But these are also fields in which pure cognitive power may be more important than experience, and rather unusual in that. For neurosurgeons, journalists, CEOs and artists – and indeed for most of us – it is highly unlikely that we will produce our best work in the first decade of our career. In fact, there is evidence that for many skills the biggest difference between those who are the very best, as opposed to just pretty successful, is simply the extra hours of practice – that is to say, the extra experience – they have had.

With a combination of experience and biological programming, our brains change significantly throughout the course of our life. From the helpless immaturity of a newborn baby to the helpless senility of someone in the last days of dementia, each function of our brain undergoes a characteristic profile of development and then, unless something else kills us first, declines. How early do we develop a brain that is 'good enough', and when do we actually peak? In this section we consider how the brain grows and develops, when it peaks in function, and for how long it is 'good enough' for what we need.

The first months and years: why can't human babies do much?

At birth, a human baby is a pretty helpless creature. Unlike the young of most other species, our babies cannot hide themselves from danger, let alone run away from it. They cannot stay warm or find nourishment. In fact, they can hardly contribute to their own survival in any way. As a result of the trade-off between the shape of a pelvis that supports a grown woman walking upright, and the maximum head size that can be fitted at birth through such a pelvis, human babies are born with less-developed bodies and less-developed brains than even our closest relatives in the animal kingdom.

That newborn baby will remain unable to survive alone for a long time. But as any proud parent will testify, she will make enormous advances in mental abilities in the first few years of life. Consider language: within the first year she will spontaneously start to parse the stream of noise uttered by those around her into distinct words, then figure out the meaning of each, and learn how they fit together. Within a couple of years she will routinely understand, and start to use, multiple new words each day. In just a few more years, she will learn to use complex symbolic systems such as drawings, letters and numbers to convey ideas and systematise the world around her. This is a feat that no other animal has ever managed. From such a biologically disadvantaged start, the rate of development of the human brain and its functions in the first few years of life is really very impressive.

Since we can't directly observe most of this rapid brain development, the first signs that we notice are new-found physical and sensory abilities: the ability to make sounds, to recognise and reach for objects, the first smile. These milestones occur in a fairly predictable sequence, so whilst some children take their first steps before uttering their first words, and others the opposite, all children learn to walk before they run, and to utter single words before full sentences. These milestones are important markers, evidence that, like muscles, brain circuits are developing correctly and becoming increasingly effective. And at the earliest stages of life it is easier to infer that normal brain development is occurring by asking parents about whether their child has taken her first steps or uttered her first word than by trying to assess the mental life of the child.

Because there is a great deal of variability within the range of normal development, this means we can't say anything accurate about whether little Angela, who learned to walk at just ten months, is destined for greatness. Although we can't make predictions at the level of an individual, we can say that, on average, children who reach developmental milestones earlier within the normal range tend to have 'better' brain structures (such as increased amounts of grey matter) and higher cognitive test scores, even many decades later. And some of those who lag behind in childhood milestones are among those most likely to show other signs of atypical development, and to be eventually diagnosed with disorders that we understand to reflect aberrant neurodevelopment, including autism and schizophrenia.

As well as suggesting that brain development is on track, reaching early-life milestones also allows babies to considerably ramp up the rate at which they can learn about the world around them. Even before crawling, the more control you have over your fingers and arms, the larger the number of things that are accessible to touch, and drop, and taste. So developing motor skills more quickly helps you develop other sensory and cognitive skills, in a virtuous cycle known as a developmental cascade. This is the idea behind many well-meaning parental purchases of so-called educational toys: providing interesting objects that are supposed to increase a baby's stimulation will in turn help speed their brain development.

Interestingly, one recent US study, which tested whether it was really possible to intervene in young babies' development like this, concluded that it could well be. A group of psychologists recruited three-month-old babies who were not yet able to successfully grasp nearby objects, a skill that usually emerges between four and six months. They then gave half the babies a two-week training experience where they used Velcro-covered sticky mittens to help them grab hold of toys they were reaching for. The children who had the mittens, and consequently experienced success at getting hold of toys, were more likely to try grasping at objects even without the mittens. Further to this, a year later they showed more advanced motor exploration and attention skills that those whose mittens had been Velcro-less. So giving the infants more ability and incentive to explore new objects at such an early stage in life seemed to have long-lasting benefits for brain development.

What is normal for a brain anyway?

Let's get technical for a moment. What does normal human brain development actually consist of? Well, to start with it's just the same as all other vertebrates. The brain begins developing around the third week after conception. Cells start off undifferentiated but gradually change into different types and eventually form into complex structures. The process is controlled by signalling molecules that tell the undifferentiated cells where to go and what to become. In the first few weeks of pregnancy, the brain grows into a smooth tube-like structure with bulbous swellings that form into the three gross divisions of the brain (the forebrain, midbrain and hindbrain). Around weeks seven and eight, neurons start to be produced and distinct brain structures to emerge, including the gyri and sulci (these are respectively the ridges and grooves that are visible on the outside of the brain; they are the way we pack the very large outer layer of cortex into a smallish skull).

Fast-forward to the first few years of life, and the human brain is continuing to change considerably. One particularly important change is the number of connections between neurons. In babies, the number of connections is actually much higher than in adults, and a lot of what happens in the brain during adolescence is a gradual pruning of connections. You can think of this as a process of increasing efficiency: connections that are not needed are weakened or removed, while those that are used often are strengthened, resulting in a leaner, more efficient signalling system.

Another thing that increases how quickly and efficiently a signal can pass through different pathways in the cortex is myelination of the axons, the long thin parts of the neurons that conduct electrical signals to other cells. If you remember from Chapter 1, myelin is a fatty substance that coats the axons, acting like an electrical insulating tape to help signals be passed as quickly as possible down the axon. This myelination doesn't all happen at once. The order in which it occurs makes sense from the point of view of how new functions come online in the developing child. Broadly speaking, pathways that deal with processing sensory information such as visual and auditory stimuli are myelinated first; then motor pathways, which deal with movement; and then cortical association pathways, which deal with the integration of information and higher-level cognitive processing, last. Geographically, myelination starts in the spinal cord and brain stem and moves gradually towards the front regions of the brain. Early myelination starts before birth but most happens in the first few years of life.

The last pathways to be myelinated are those in the prefrontal cortex, the very frontmost section of the brain. This is the area that deals with the highest-level cognitive functions; the area that has expanded most recently in our evolutionary history and that most distinguishes human brains from other species'. Unlike most brain areas, the prefrontal cortex doesn't mature until early adulthood, with myelination of pathways in this area not complete until the late teens or early twenties. This reflects the fact that advanced cognitive functions that depend on this brain area are also still developing. These are functions like working memory – the ability to hold information in mind while you act on it – and aspects of attentional control, and executive functions such as the ability to select and switch rapidly between competing demands on your attention. Directly reflecting the slow maturation of this part of the brain, these functions reach a peak in most people as late as their third decade of life.

How do different cognitive skills develop – and why?

How the brain develops gives us clues as to the order in which different aspects of cognitive function would be expected to progress. It is possible to measure the cognitive development of the child from surprisingly early in life. Whilst babies do not have the words to tell us much about their internal life, there are a number of tricks that developmental psychologists use to reliably measure cognitive differences between infants, and changes in abilities as a baby develops. One example is preferential looking. Babies naturally look more at things that are new to them; they are programmed to find unfamiliar things more interesting, which is a sensible strategy when you are trying to learn about the world. Psychologists exploit this tendency by putting two items in the baby's line of sight and then measuring which she spends more time looking at. In this way, it is possible to objectively test whether a baby can, for example, tell the difference between a photo of a familiar person versus a person she has never seen before, or whether she remembers an object that she was introduced to a few minutes earlier. So abilities such as distinguishing shapes, objects and sounds, and remembering them, can be tested in this way from very early on in life.

The earliest real attempt to describe human cognitive development more broadly was produced by a Swiss psychologist named Jean Piaget, in the first half of the twentieth century. Piaget thought there were four major stages of cognitive development, with sharp changes marking the shifts in them. In the first, 'sensorimotor' stage, infants don't yet have language, and they learn about the world from physically interacting with it. In the 'pre-operational' stage, from around two to seven years, children form stable concepts, and begin to reason and wonder why things are the way they are. However, they find it hard to understand anything other than their own view of the world and as such their logic is often flawed. During the 'concrete operational' stage (around seven to eleven years), their ability to reason about real (concrete) events becomes fully formed, but they remain unable to reason correctly about hypothetical events. This last ability is developed during the 'formal operational' stage, which is characterised by new abilities such as abstract thought and metacognition (thinking about thinking).

Piaget's theories were ground-breaking in their description of a sequence of cognitive stages of increased sophistication. The stages are still considered a reasonable approximation of the way that children develop, though subsequent experiments and modern neuroscience have allowed us to better tease apart different aspects of development in different aspects of mental function. In particular, since Piaget's time we've learned a lot more about how best to categorise the different cognitive functions, and what brain networks and regions each depends on.

One way in which Piaget's theories were novel was in describing the child as an active learner who updates her understanding of the world as a result of learning new things about it. This sounds sensible but raises the question: what actually limits the rate of that new learning? Is it just that a certain number of experiences and opportunities have to be accrued before the next stage can be reached? Or is a child's ability to learn ultimately limited by the size or maturity, or some other kind of fitness for purpose, of the structures and connections in the developing brain?

To answer this, let's use the example of how language, that most human skill of all, develops. In most cultures, babies are exposed to words right from birth. In fact, there is some evidence that by indulging in the 'cootchy-cootchy-coo' kind of baby talk we are instinctively providing babies with lots of the sort of stimuli that help the very first stages of language-learning. It is obvious that the language you end up learning is driven entirely by the environment that you experience; that is to say, we humans are good at learning languages but are not innately programmed to learn any specific language.

We also know that there are critical periods during which the brain seems particularly sensitive to some aspects of language. For example, whilst all babies can tell the difference between the sounds R and L, those adults who are never exposed to them when young (for example many Japanese adults) cannot. So there is thought to be a period of brain plasticity during which such subtle differences in sounds need to be heard if they are later to be detectable. From unfortunate cases of children raised in neglectful, isolated or abusive circumstances, it seems that it is difficult to learn to fluently use the grammar of any language if you are not exposed to some form of grammar before puberty. And, as many of us can testify, it becomes more difficult to learn a second language, and much less likely that you will ever become truly proficient in it, the later in life you start.

So there seems to be a biological driver for language-learning, or at least a period in which the human brain is particularly primed to develop it. This may be related to the general fact that brain plasticity is greater earlier in life; perhaps language-learning is especially dependent on this. Alternatively, it might be to do with the relatively specialised neural structures by which language is supported in the brain.

When you look at a brain from above it appears pretty symmetrical, with two halves, which we call hemispheres, and a deep groove running in between the two. During development, many functions of the brain specialise into one or other hemisphere. For about 90 per cent of right-handed people, and about 50 per cent of left-handers, it is the left hemisphere that becomes dominant for language. What this means is that two important language regions, Broca's area in the frontal lobe and Wernicke's area towards the rear of the brain, develop in the dominant hemisphere. Damage to these two areas produces characteristic problems with different aspects of language. Typically, people with damage to their Broca's area have problems producing language, whilst those with damage in their Wernicke's area have problems understanding it.

Language is a great example of how complex and specialised the human brain is. But, as a quick read of Stephen Pinker's brilliant book *The Language Instinct* reminds us, language really is a human speciality and a peculiarity. It's therefore not a good example of how a typical brain function develops. To answer our question about brain primetime, we really need to consider a more normal brain function.

A lifetime of learning

If language is not such a good exemplar for understanding how much brain we need to develop, let's instead consider learning and memory. There are many different forms of memory that psychologists talk about, and we can divide them up any number of ways. For example, timescale: short-term memories might be those lasting seconds or minutes, whereas long-term memories could last a lifetime. Or the different ways we use memory: episodic memory, for example, is what you use when remembering events you witnessed (what, where, who), whilst procedural memory is what you are using when you practise an already-mastered skill, such as tying a shoelace or riding a bike.

For most forms of memory, humans are not at the top of the evolutionary tree. Associative learning, i.e. understanding that there is an association between two particular events or stimuli, is something that many species excel at. Think of Pavlov's dogs salivating when they heard the bell, or even an optimistic Labrador that has a very clear understanding that someone putting their shoes on immediately after breakfast is about to go walkies. Many other species have been demonstrated to be excellent at this form of learning, such as pigeons, which can be taught quite easily to distinguish between complex shapes, if there is a snack in it for them. So too are even very young children, as anyone who has ever played a six-year-old at pelmanism (the game where you have to remember the location of matching pairs of cards) can testify. In fact, this is a skill that develops, and peaks, early on in life in humans, with little difference in skill noticeable between six-year-olds and thirty-six-year-olds. So if pigeons, young children and Labradors have all mastered the ability to learn associations between stimuli and rewards, should we infer that this form of memory requires a less developed brain, or a less complex brain network, than, say, language?

One answer to this comes from consideration of what happens later on in life. Beyond about forty years old, the ability to form new associations starts to decline, first gradually and then at a rapidly accelerating rate. Forms of memory that depend heavily on the hippocampus seem particularly affected, such as those that involve associating objects and locations ('Where did I leave my keys last night?'). These types of memory failure are characteristic of the early stages of Alzheimer's disease, which usually begins in this area of the brain, but are increasingly common in healthy ageing also. If we think of a graph of associative-learning skill in humans then it would peak early, remain flat for around thirty-five years, and then start an accelerating decline from the fifth decade.

In sharp contrast, aspects of language use clearly improve with time. If we plotted a graph of the size of your vocabulary throughout life it would look quite different. We learn new words fastest during our early years but, in general, the size of our vocabulary continues to increase as we are exposed to more and more words: in the conversations we have, the books we read and the radio and TV we listen to. In English, unlike some languages, there is not always a one-to-one relationship or direct correlation between how words are spelled and how they are pronounced. So one way of assessing how big a vocabulary someone has is to give them a list of irregularly pronounced words (words like 'yacht', 'cough' and 'though') and ask them to read it out loud. Clearly this is testing one particular kind of memory, so you might expect it to decline with age as other forms of memory do. But this form of memory is spared by both normal ageing processes and many neurodegenerative diseases. In fact, it is so noticeably spared that it has become the standard way psychologists and neurologists estimate a person's previous level of intellectual functioning, when assessing the extent of brain damage that has occurred after a head injury, or in a person with dementia.

Adult brain function – when does all the downhill start?

Even through the middle of life, when you might think we are neither advancing nor in rapid decline, the brain is not a static organ. Given how early our fertility peaks, it is interesting to note that some aspects of brain function are still developing into perhaps our mid-twenties. Other things going on in the brain that, on the whole, we believe are healthy or helpful continue even later. For example, the amount of white matter increases until mid-adulthood, then remains essentially static while we gradually lose grey matter. The cortex starts to thin out from early childhood, but this accelerates rapidly from the age of about fifty-five, and older brains also have narrower ridges and wider grooves. In other words, our brains shrink: they get lighter, and smaller in volume, and the ventricles (cavities filled with cerebrospinal fluid) get larger. In our twenties, a male brain weighs around 1.4 kilograms and a woman's around 1.3. Brain shrinkage accelerates from our forties onwards. By our mid-sixties, a male brain weighs around 1.3 kg, and by age ninety, around 1.2 kg.

So much for the brain at the macro level. What about at the micro level? Well, it's not that rosy there either. The neurons in older brains are smaller and they have a simpler network of connections between them. Even in healthy older adults, it is common, indeed normal, to see signs of damage such as a build-up of the protein 'plaques' that are associated with Alzheimer's disease, and microscopic bleeds that are signs of vascular damage.

Some structures suffer greater loss of cells than others. The hippocampus, for example, shows considerable age-related change, even among those who do not have dementia. The frontal lobes, especially the prefrontal cortex, seem to show increased cell loss and reduced connectivity within and between areas. The functional consequences of these changes presumably underlie what we think of as 'normal' cognitive ageing: the gradual worsening of memory, for example, that we notice from middle age.

You might think it would be easy to work out the trajectory of each cognitive function with age. It's actually surprisingly difficult. The first challenge is that, in trying to see how functions change over time, you have to do one of two things, both of which have flaws. First, you could take a young child and then give them a comprehensive test of their cognitive abilities, say every year for eighty-five years, and see how their results change with time. This is difficult, and expensive, and so requires scientists and their funders to show a great deal of foresight and patience. There are a couple of scientific challenges too. For example, if you're interested in maths ability, how do you measure it in people of all different ages? You clearly can't ask a five-year-old the same maths questions as an eighteen-year-old. You might want to ask a five-year-old to do a simple adding sum, and ask an eighteen-year-old old to solve a differential equation. But it's hard to know whether these are really tapping the same core ability, and it's quite possible that whilst some children who learn to add up easily go on to be excellent at algebra, others never really master it. So following people longitudinally works best where there is a core skill that can be measured on the same metaphorical ruler at each age, such as pelmanism, or the speed of finger-tapping.

For skills that can be measured in the same way across the lifespan, you can take a faster route to work out an age profile, by asking people of all different ages to do the same test and plotting average scores by age, rather than following individuals for their entire life. It's easier to get people to do a cognitive test once than to get them to come back every year for eighty years, so you can probably persuade a more representative sample of the population to take part. And if everyone only takes the test once, their scores won't be affected by factors like how much you benefit from practice. However, there is a downside to these cross-sectional studies. Today's five-year-old has already had a rather different experience of life than the eighty-year-old did when he or she was five. *That* five-year-old was living through a war. Food was rationed; his father might be absent, or dead. There was no TV, no iPad, no Baby Einstein. Tuberculosis and polio were still rife. By making it through to our (fictional) study, that five-year-old survived far greater risks to life and limb than today's five-year-olds face. Conversely, today's five-year-old has other concerns, which may or may not have had some effect on brain development. He has, on average, fewer siblings to learn from or compete with. His mother and father are far less likely to be married or living together. He has a far greater risk of being diagnosed with autism, ADHD, asthma or food allergies, and is far more likely to be overweight.

This matters because it means that when we compare an eighty-year-old with a five-year-old today, we are measuring not just a difference in age but also a difference in many environmental factors. A boy born in 1935 had a life expectancy of around sixty; one born today can expect roughly twenty years more. The mean height of adult men has increased at a rate of approximately one centimetre per decade for the last 150 years or so. These increases in life expectancy and height are thought to reflect welcome reductions in exposure to disease and improved nutrition, especially during childhood.

Perhaps unsurprisingly, the same trends of improvements across generations are seen in brain function. Intelligence quotient (IQ) is a standard way of expressing a person's general cognitive ability and is defined on a scale with a mean score of 100 points. There are plenty of interesting arguments to be had about what intelligence really means and how best to measure it, but most IQ tests involve sitting down with a pencil and paper for a couple of hours and completing tasks of verbal and numerical skills and abstract reasoning. The most common tests, like the Wechsler adult intelligence scale (WAIS), involve ten or more separate tests, with the scores added together and standardised according to age to give your overall IQ. The interesting thing about this from our point of view is that the average test scores of the whole population have been getting better in every generation. Our best guess suggests that IQ has gone up in Europe and America over the past century at rates of around three points every decade. This is enormous: it means that someone who fifty years ago was Mr Average, with exactly half the population better and half the population worse than him, would find that 84 per cent of today's population was better than him. Some of this improvement in IQ may be due to non-brain-related factors, such as increased familiarity with the tests, but taken in conjunction with the parallel changes in how we are advancing physically, we can assume at least part of it is due to having stronger, more efficient, or possibly just plain bigger, brains.

Cognitive peaks and troughs

So in asking how different brain functions change with age, we are in part asking what changes in an individual's brain, and in part asking what's changing with time in the human brain in general. We will come back to this latter point in subsequent chapters. For now, let's return to our original question in this chapter: when in our lifetimes can we expect our cognitive peak to be?

Possibly the biggest ever attempt to answer this question was recently published by two Harvard University researchers, Joshua Hartshorne and Laura Germine. They obtained cognitive scores of various kinds from nearly 50,000 volunteers, and looked at the age group in which performance peaked for every function they could find.

As we would expect, performance on tasks that benefit from experience, or from gradually accrued information, peaked relatively late in life. For example, tests of vocabulary and general knowledge were performed best by people of around fifty years old. In contrast, tests in which the youngest adults seemed to do best might be thought of as those benefiting from raw processing power over experience: measures that require short-term memory, the ability to switch rapidly between tasks, tests of abstract reasoning. All of these peaked, in this sample, in the early twenties. So given the difference in their fields of expertise, Einstein's intellectual capacity may well have been at its peak in his twenties, but Shakespeare, who supposedly died on his fifty-second birthday, could have been cognitively improving where it mattered right up until the end of life.

What happens beyond the fifties? we hear you ask. In this study at least, every cognitive function, without exception, worsened beyond that age. But interestingly, in a second sample, which was recruited and tested through the internet, vocabulary wasn't seen to peak until around sixty-five years old. One possible reason for the difference is that these silver surfers represent an empowered group of older adults who continue to take part in intellectually stimulating, knowledge-expanding activities into later life.

On 8 November 2016 the US population went to the ballot to elect their next president: arguably the most powerful job on earth. The choice for this world-leading role was between the sixty-nine-year old Hillary Clinton and the seventy-year old Donald Trump; when the latter won, he was the oldest US president ever to be elected. It is clear from the evidence above that essentially all cognitive functions pass their peak long before this age. Yet the highest ranks of world politics, like most industries, are dominated by people who are, by these measures, well past their best. Why is this? One possibility is that, whilst cognitive function slows down in later life, this is outweighed by the benefits of having more experience in a particular role or industry. Perhaps you are more likely to make the optimal decision when you have more in-depth knowledge, or more chance of having previously encountered a similar situation. Indeed, as their raw cognitive skills decline, older workers may be forced to adopt a strategy that leans more heavily on reusing knowledge gained from prior experience.

In the technically and socially complex worlds in which most of us spend our working hours, it is hard to measure whether we change strategy in order to keep up as we age. But in more controlled environments we can look for evidence that as we get older we use different cognitive strategies, and sometimes different brain networks, to get the same job done. In another web-based study, this one including more than 10,000 people aged 10–70 years old, participants spent four minutes watching as black-and-white photos were shown, with gradual transitions so that each photo melted into the next. The participants' job was to judge whether the photo was of a city scene or a mountain scene, a decision that they indicated with a key-press. From this very simple test, the researchers calculated four different measures of cognitive function. Two of these reflect how good a participant is at the task in hand: broadly speaking how accurate and consistent they were in responding to the pictures. The other two measures reflect not skill at the task, but something more like the attitude the participant was taking towards it; how quickly they responded when a picture started to change, and how willing they were to respond in the presence of uncertainty. The latter two measures give us an idea about how individuals might differ in their strategies on this kind of task: would you press as soon as you thought you perceived a slight change, or would you wait until you were really sure?

What effect did age have on these different aspects? Well, the two measures that reflected ability to do the task showed a typical performance curve across the lifespan. Performance improved rapidly between ages ten and sixteen, then much more slowly from sixteen to the middle forties, after which it started to worsen. But for the two measures related to *how* the task was tackled, the picture was quite different, with the most 'risky' or 'impulsive' responding – i.e. being more likely to press when they thought they saw a change – seen in people in their mid-teens. After this age, behaviour followed a linear, increasingly conservative trend, with responses becoming less impulsive as the age of the participants increased.

Completing an online cognitive test is probably not going to be the most exciting four minutes of your day, but nor is it likely to be the most challenging. For most of us, the challenges we face at work require not our raw intellect, but rather our ability to apply it in complex social environments, where other abilities – to influence, cajole, understand subtext, manage expectations and reach consensus – are also key. So a second possible explanation as to why the most senior people in organisations tend to be older is that, through a lifetime of use, they have learned to apply emotional or social skills to greater effect than their younger colleagues.

Social and emotional skills are very interesting aspects of brain function, which develop relatively independently from the core cognitive domains we have considered so far. Babies express emotions in their faces in ways that are recognisable regardless of the culture they are brought up in, so we think that this is an innate behaviour. By three months, babies can tell the difference between happy, surprised and angry faces, an impressive feat given how immature the visual system is at birth. At around one year old, children start to use other people's facial expressions as cues in interpreting the significance of events going on around them.

There is a difference, however, between recognising other people's emotions and being able to deal with, or control, your own, and in this latter aspect, older adults are thought to do particularly well. Work by Susan Charles and Laura Carstensen, two Californian researchers, suggests that older adults may also be more attuned to the emotional aspects of tasks whilst younger adults often ignore them, showing less activity in emotion-related brain networks. As we get older, the 'pure' (non-emotional) cognitive networks get less efficient. This may make it more difficult to ignore or inhibit emotional aspects of information or decisions – it may be harder, say, to turn down an offer for something you don't want, because you are more sensitive to the feelings of the person making it. Alternatively, they suggest, perhaps forcing information to be processed at a slower pace inherently gives more time for reflection and taking into account the emotional aspects of an issue, leading to what may be 'wiser' solutions.

The lifetime of a brain

It's clear that some aspects of brain function are optimised within the first few years of life, whereas others emerge more slowly and improve over the course of many decades. The pharaoh Tutankhamun was nine when he ascended the throne, and his nine-year reign is considered highly successful. We may not necessarily want pre-adolescents running our countries, or indeed our companies, in the modern day. But up until very recently in human history, life expectancy and old age would have been so constrained by factors like food availability, disease and physical frailty that the late-life functioning of the brain would have had a relatively small impact on either survival or quality of life. Now that we live, and especially work, far longer, a better understanding of the peaks and troughs of different brain functions across life may be key to understanding how to enjoy a well-rounded intellectual lifetime. The raw cognitive processing power of early adulthood may not be sustainable through life, but rather than simply mourning its loss as we age, we should enjoy the benefits that accumulated knowledge and changes in processing style can bring. Leaving aside the first few years of childhood, our brains are probably good enough for us to survive and thrive from surprisingly early on in life. And if we're lucky, we can usually remain that way for at least the biblical 'threescore years and ten', despite the brain's gradual slowing and shrinking.

So now we know something about how our brains change with age; but age isn't the only thing that can bring about variations in our cognitive functioning. Let's now explore the impact of other influences on our brain.

CHAPTER 5

Good Days and Bad Days: How does brain function vary from one moment to the next?

According to her fancy new smartwatch, last night Sally slept for 7 hours and 23 minutes, of which 3 hours and 7 minutes were deep (REM) sleep. She ran 5 kilometres (3.1 miles) this morning before breakfast, which comprised 450 calories including 18 grams of protein. She had her second coffee of the day twenty minutes ago. Sitting down to open her computer and start the day, Sally feels like she is at her cognitive peak. Is she?

In the last chapter we discussed how the maturation of the human brain over a lifetime results in a shifting pattern of optimal functioning, with changes in brain structure at the macro and micro levels that follow a gradual process of development and decline throughout life. So we already know that the brain is a dynamic organ. But just how dynamic is it? When we feel brain-dead at the end of a long meeting, or cognitively impaired the morning after a big night out, does that really reflect a measurable difference in how well our brain is working? In this chapter we explore how variable normal brain function is, and how the things we do, and don't do, drive the peaks and troughs of brain function from year to year and from moment to moment.

Measuring brain changes

To understand how much brain function – any brain function – varies from moment to moment, you have to be able to measure it. As you'll have realised already, measuring activity in the live human brain is not the easiest thing to do. For one thing, the brain is well protected from the outside world, mostly by the skull but also by some sinews, skin and varying amounts of hair. So we cannot directly watch or touch it. We also cannot, with rare exceptions, risk damaging it by inserting scientific instruments that directly contact it. Instead, we've developed a number of techniques that allow us to measure aspects of brain function in non-invasive ways.

We call techniques that allow us to build up pictures of the structure of the brain (or of the way that its constituents, such as blood, move around) neuroimaging. You'll probably have seen pictures from CT or MRI scans, which are a bit like X-rays but optimised to show up the differences between types of tissue of varying degrees of squishiness, rather than hard bones. They provide a snapshot of the structure of the brain, and can be very useful for measuring major changes, such as the shrinkage caused by ageing, or damage caused by events like a stroke, tumour or head injury.

If we take repeated snapshots of the brain over time, and then give the data from those images to some very clever statistical folk to process, it's possible to build up a picture of what is changing in the brain over a given period. This is the basis of something called functional MRI (usually shortened to fMRI), where many hundreds of MRI images are taken over the course of several minutes, in order to measure how blood flow is changing in the brain. To be slightly more specific, fMRI measures how blood (in fact, the oxygen that is carried by the haemoglobin in the blood) is being moved to different areas of the brain. We assume that the areas that need more oxygen are working harder at that moment in time. So, for instance, by comparing fMRI images taken during two different tasks – say, an easy and a more challenging mathematical problem – we can work out which areas of the brain work harder, or which extra areas of the brain are needed, in order to solve a more difficult problem.

Some other ways of measuring brain function don't require you to go into a scanner and so can be more useful where you don't happen to have access to these enormous and hugely expensive pieces of kit. One common way is an electroencephalogram (EEG). While beautiful, cool-looking EEG devices are now being built by Silicon Valley start-ups, the standard scientific-grade EEG machine normally consists of a thing that looks like a skullcap with a load of wires sticking out of it. Built into the cap are a few dozen electrodes, which sit directly on the scalp, sometimes with some goo squeezed in to maintain a better connection between the scalp and the electrode. The idea of EEG is that it measures the location and speed of waves of electrical activity as they pass over the head. These waves reflect the firing of neurons in different areas of the brain. Because of all the bone and flesh that is in the way between the brain and the machine, this isn't a very spatially precise way to measure brain activity, unlike MRI. But, because electrical activity moves more quickly than blood does, EEG data reflects the timing of brain activity with higher precision than MRI.

There is a third way to measure brain activity that, while being less direct than either fMRI or EEG, is often cheaper and easier, and that is running psychology experiments. In these, volunteers are asked to complete a task or puzzle, which has often been refined and standardised over decades of use. This long history of use means that we know a lot about the neuroscience behind each task: which brain regions or networks are needed for successful completion, how performance on the task is typically affected by factors like age, gender and educational level, whether patients with a given disorder show specific patterns of impairment on the task, and whether a particular drug is likely to improve or worsen performance.

One commonly used task is the Tower of Hanoi or Tower of London. In its original form this looks like a simple children's game. Wooden discs of varying diameters are stacked in size order on one of three rods, so that the smallest disc is at the top and the largest at the bottom (see Figure 5). The aim of the game is to move all of the discs on to one of the other rods while obeying three rules: that only one disc can be moved at a time; that a smaller disc can never be underneath a larger one; and that only the topmost disc of a stack can be moved. If there are only three discs this task is relatively easy and can be solved by making a minimum of seven moves. But to successfully complete the game with seven or eight discs in the stack is much more difficult.

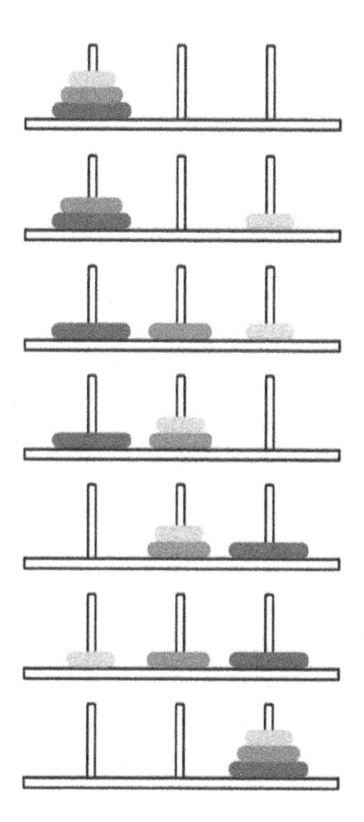

Figure 5: The Tower of Hanoi: The minimum number of moves to solve a three-disc puzzle is seven

The Tower of Hanoi was originally a mathematical problem, invented by the French mathematician Edouard Lucas in 1883. It can be systematically solved by following a mathematical strategy, and is always solvable in 2^n-1 moves, where *n* is the number of discs. In the 1980s the Tower of Hanoi was adapted by London-based researcher Tim Shallice into the Tower of London: a test based on the same principles as the Tower of Hanoi, but giving the volunteer different goals each time, so that it could be used repeatedly to test certain aspects of cognition. Shallice reported that people who had received brain injuries affecting their frontal lobes found it particularly difficult to complete the task, and subsequent studies, which ask healthy volunteers to complete the task when in a brain scanner, have confirmed that doing it activates parts of the prefrontal cortex. For most non-mathematicians, successful completion of the more difficult levels requires you to develop a plan or strategy and then keep a close track of what you're doing using working memory as you carry that plan out. As these are key functions of the prefrontal cortex, it's no surprise that people do best at this task when they have an optimally functioning prefrontal cortex, and that children and old people, and those with disorders such as schizophrenia and ADHD (which affect this region), find it harder to do.

You'll notice that brain activity is not measured directly here, but is inferred based on what is known from previous research, on the basis of the answers or responses that the volunteer makes. Although not nearly as much fun as fiddling with really expensive kit, well-designed psychology experiments can give us surprisingly precise answers about how brain function varies from moment to moment, as well as over longer periods.

There's one last catch. The techniques described above are all very good at telling you what happens in an average brain, or at least the brain of the average person who agrees to take part in psychology studies. But if you want to learn about the variability of human brains, you need to study larger numbers of people. The very largest studies are often called epidemiology studies (from the same root as 'epidemic', meaning things that are prevalent among the people). In an ideal scenario, these involve the study of the entire population, so that conclusions can be drawn not just about the brain of the average inhabitant of a country, but also about how much variation there is amongst her compatriots. While we don't have the time or money to put entire populations in scanners, we can sometimes exploit other data that do exist for entire populations, such as medical or school records, accident statistics and, in some countries, data collected from people conscripted into military service.

So when we're looking to understand an effect on the brain, we are often selecting between techniques that give highly detailed data about very small numbers of people, or superficial data about large numbers. If we're lucky, these different ways of approaching the problem give us converging evidence.

Enough about the technicalities. What do we actually know about the state of your brain, right now, as you read these words?

Seasons of neural life

As well as the one-way process that is ageing, some other long-term trends are likely to be affecting your brain right now. Imagine, if you will, that as you read this you are sitting outdoors in late spring, on one of the first days on which it is sunny enough to be outside without a sweater. Assuming you don't live on the equator, for the next few months the hours of daylight will be longer, the average temperature higher, and the amount of UV light you are exposed to considerably greater than it has been through the winter. Perhaps you will more often cycle or walk places in the coming months, rather than driving. You may spend more time outdoors in parks and the garden, and with the windows open, and consequently share less recycled air and colds and other viruses with friends, family and workmates. It's fair to say that many of us feel healthier and happier in the summer months. But is this directly due to our brains?

'Winter blues', or seasonal affective disorder (SAD) as the clinical condition is called, is pretty common among people living outside of the tropics. It's likely that the vast majority of us experience some of its symptoms. In a study conducted in Maryland, USA, for example, 92 per cent of people reported that they noticed seasonal changes in their mood and behaviour; 27 per cent found these problematic, and somewhere between 4 and 10 per cent of people met criteria for SAD.

SAD is just one of several brain disorders that appear to be tied in with the seasons. Studies from around the world of people with bipolar disorder (sometimes called manic depression) show peaks in manic episodes in spring and summer, and in depressive episodes in early winter. One reason for this may be seasonal changes in sleep patterns, which in some people seem to be an early warning sign of an upcoming episode of mania or depression, and in others seem to actually trigger a change in mood.

Epidemiological studies have suggested that the seasons affect risk for other brain-related disorders too. It has been known for a while that babies who are born in the winter and spring are at higher risk for schizophrenia than those born in summer and autumn. It's not completely clear what causes this, but it is likely linked to exposure to seasonal risks during the neurologically crucial third trimester of pregnancy. One such risk would be maternal influenza, or other infections. A second would be vitamin D deficiency, which has also been implicated in other brain disorders, such as autism and multiple sclerosis. Some vitamin D can be gleaned from the diet, but the majority is generated by exposing our skin to sunlight. As we collectively spend more time indoors, use sunscreen and worry more about skin cancer and the effects of sun exposure on our wrinkles, vitamin D levels in many populations are dropping, and in countries where strong sunlight is not available year-round this can cause significant deficiencies during the winter months. Indeed, vitamin D deficiencies are common even in some countries where sunlight is available year-round: one study carried out in a hospital in the United Arab Emirates found low levels in 86 per cent of blood samples taken from UAE nationals and 79 per cent from non-UAE nationals, with extreme deficiencies in 28 per cent of UAE nationals and 18 per cent of visitors. Whilst cultural dress and customs may play a part in reducing sun exposure in the UAE, low vitamin D levels are common even in countries where these factors may be less influential; in Australia, for example, it is estimated that almost a third of adults have low levels.

Large-scale studies tell us something interesting about seasonal influences on brain development and long-term health. But even for those of us who live in regions far from the equator, seasonal differences in behaviour are now relatively subtle. If we split an average workday into three parts, work, leisure and sleep, we see relatively little difference in how we spend our hours in summer versus winter. We probably spend our free time somewhat differently on balmy July evenings from how we do on freezing January ones. But at least for people with indoor-based occupations, the hours spent at work are not greatly affected by the season, nor is sleep pattern. Modern technologies, from central heating to waterproof clothing to reliable transportation, mean that our lifestyles are probably dramatically less affected by season than those of our ancestors. Supermarkets with global supply chains mean that while you may notice, for example, how nice strawberries taste when ripened under a local sun, you can pretty much eat them all year round if you want to.

This reliable season-free nutritional landscape is not only quite different from that of our ancestors, but also from that of all wild-living creatures. For most species, variation in the availability of food dictates huge seasonal changes in behaviour – think of the annual migration of the whale, wildebeest or swallow. And for almost all species, food availability is thought to underpin the timing of reproductive cycles, with all the associated nest-building, antler-locking and colour-changing that go with it. Human seasonal changes in behaviour and mood look quite small in comparison. So if we are interested in understanding what happens in the brain as the seasons wax and wane, a good place to start is to look at what happens in species that may have smaller brains than us, but that have developed far greater changes in seasonal behaviour.

Follow the birds

In their songs, migrations, plumage and nesting, birds epitomise seasonal changes; we traditionally note the first swallow we see as the start of summer. By comparing species of bird that show different seasonal behaviours but are otherwise similar, we can make inferences, or at least good guesses, about the relationship between these behaviours and the brain.

One brain area that is particularly well studied in both birds and humans is the hippocampus. In both species, the hippocampal formation is crucial to memory, particularly spatial memory. Birds with damaged hippocampi show such problems as difficulty finding places where they have previously stored food, or navigating and recognising landmarks. Bird species who need to do more spatial navigation have a larger hippocampal area than similar birds who do less. For example, homing pigeons have a larger hippocampus than non-homing strains, and migratory subspecies of birds tend to have larger hippocampi than their non-migratory cousins. But do these size differences reflect evolutionary adaptations – that is, genetically programmed species-wide effects – or just the effect of each individual bird using its hippocampus a great deal?

The structure of our brain adapts across our lifetime in the light of the use that it is put to. In one famous example, researchers used brain scans to measure the size of the hippocampi of London taxi drivers, who as part of their training have to learn 'the Knowledge', a mental map of the whole of London, to allow them to take the best route between any two points. They found that the hippocampi increased in size as the drivers underwent the strenuous training process and learned more London landmarks. Indeed, the longer the cabbies had been working, the bigger the changes noted in their hippocampi, suggesting that years of negotiating London's streets continued to make a mark on their brain even once they had mastered the layout and passed the Knowledge. In the era of GPS, where any smartphone enables its user to become an Uber driver, would we still expect to see these occupationally driven hippocampal changes? Since the brain functions largely on the basis of 'use it or lose it', we might predict not.

In the bird world, the hippocampi of chickadees, a North American member of the tit family, have been particularly well studied. Chickadees don't migrate and have a strong tendency to store food wherever they find it in autumn and winter, as sources become rarer. In contrast, when food is abundant during spring and summer they don't seem to bother storing it. Careful studies of the size and nature of the chickadee hippocampus throughout the year has shown a seasonal pattern of changes, including an overall increase in both hippocampal and individual neuron size, peaking in about October, the start of the food-storing season.

If there were seasonal brain changes in humans as there are in birds, would these changes be a consequence, or a cause, of seasonal changes in behaviour? For example, if you tend to rise earlier and exercise more in summer, it would be difficult to know if any changes in brain function were due to these behavioural changes, or due to the direct effects of longer daylight hours and warmer temperatures. In normal life it is extremely hard to pick these interrelated behaviours apart, because we cannot control the external seasonal cues. But one recent, rather interesting study managed to do just that. A team of researchers from the University of Liège in Belgium asked twenty-eight healthy volunteers to spend four and a half days living in an environment where there were no seasonal cues at all, and where every aspect of the environment and the volunteers' behaviour was kept constant regardless of the time of the year the volunteer visited. The volunteers were tested on a number of physiological and cognitive measures across the whole year: some in winter, others in summer, spring or autumn. The clever part of the experiment is that the cueless environment acted as a sort of wash-out period, so any differences seen among tests taken in summer versus winter for example, could not have been due to any immediate differences in environment or behaviour, such as hours of sunlight or warmth on the day of testing, as these were always identical for everyone. Rather, any variability that emerged between the volunteers tested in winter and those in summer must have been attributable to physiological and cognitive effects that build up more gradually as seasons change. In essence, researchers were able to measure cognitive and physiological changes that were entirely due to season, and not due to the effects of behavioural changes that accompany the seasons.

The researchers hypothesised that people would perform better on cognitive tasks around the summer solstice and worse around the winter one. In fact, they found that cognitive scores varied little across the seasons, but fMRI scans showed that there was considerable seasonal variation in the brain networks that were being used to perform cognitive tasks. For example, in a task requiring sustained attention, there was greater activation in midsummer, and less in midwinter, in a wide range of brain areas. This suggests that the brain either has different resources available, or has to use different resources to complete the same task, in the different seasons.

Sustained attention is a relatively low-level cognitive process, and a second, more intellectually challenging cognitive task showed a different seasonal pattern, with brain-activation differences peaking around the spring and autumn equinoxes.

Much larger studies have looked at the cognitive functions of people who live inside the Arctic Circle, and therefore experience enormous annual variation in daylight hours, with twenty-four-hour darkness in the depths of winter, and summers of twenty-four-hour daylight. In general, people living at these latitudes show relatively little seasonal variation in their cognitive functions. As discussed above, it may be that brain functions change to compensate for the effects of seasons, so people who live in such extreme environments have adapted to function cognitively well regardless of season. Another explanation is that people for whom short daylight is particularly trying – who are not well adapted to this environment – are probably more likely to move away from such areas.

So we can imagine an annual cycle of brain function that may be directly driven by daylight and climate, and behaviours that vary according to it. But what about other biological or social cycles?

Biological cycles and 'baby brain'

It has been suggested that menstrual cycles affect multiple brain regions that regulate cognitive and emotional processes. There is some evidence to suggest that women do better on tasks of working memory at points of their menstrual cycle where levels of the sex hormone oestradiol are highest. Conversely, for functions that are emotionally tinged (such as the recognition of feelings and the bedding-in of emotional memories) performance within individuals seems to peak in the later parts of the menstrual cycle, when levels of progesterone are increasing.

During pregnancy, changes in sex hormones drive enormous changes in body structure and function. Many women's magazines, and indeed many women, report the phenomenon of 'baby brain', a temporary worsening of cognitive function, particularly short-term memory, during pregnancy. This is backed by quite a few research studies that compare cognitive scores, typically in a small number of pregnant women recruited from antenatal classes, to their non-pregnant friends. There are a lot of plausible reasons why pregnancy might affect brain function: it brings huge surges in the hormones progesterone and oestrogen, which are active in the brain as well as in the body, and it is often, for first-time mothers at least, a huge life change. So we'd probably not be surprised to find that pregnancy affects the brain. The question is, would you expect it to make cognitive function worse, or better?

In rodents, pregnancy is accompanied by an improvement, not worsening, of memory and cognition. Pregnant rats are better than non-pregnant ones at navigating mazes and recognising objects, and they also remain calmer in stressful situations. Because most aspects of maternal behaviour are similar across humans and rats, and controlled by the same parts of the brain, a group of Australian researchers became suspicious. Why, they reasoned, would pregnancy have opposite cognitive effects in humans and rats? They thought the best way to answer the question was to recruit a group of young women who were known to be representative of the general population, and then follow their cognitive functions through repeated testing as the women did or didn't go on to have children over the next few years. In the first eight years of the study, nearly 200 of the more than 1,000 women who started the study between the ages of twenty and twenty-four went on to have children. Four cognitive tests were taken by all the women when they joined the study, and then again four and eight years later. These cognitive tests coincided with a pregnancy in seventy-six of the women.

Using these multiple waves of assessment, and a large and representative sample of women, the researchers were able to look more carefully at the effects of pregnancy and motherhood on cognitive function than previous studies had done. They found no evidence that having a baby affects cognitive function: there were no differences in any of the tests between women who did and did not have children in these eight years. There were also no substantial differences in scores in the women who took the cognitive tests while pregnant, compared with the rest of the group. One possible interpretation of the differences between this and previous studies is that the problems that some pregnant women report are due to other things that commonly occur during pregnancy, like sleep deprivation or anxiety. These may well affect memory, and it may be that the pregnant women most often recruited into the previous studies were more likely than their non-pregnant friends to be suffering from these issues.

Another possibility is that, in measuring only a few aspects of cognitive performance, the researchers missed other aspects of brain function that are more likely to change during pregnancy. It might be that memory is the most evolutionarily useful cognitive function to a pregnant rat, but might not be the most important skill to a pregnant human. When preparing to care for a brand-new helpless human, memory may be less important than, say, being super-sensitive to the changing needs and emotions of another being.

The ability to understand another person's facial expressions, read their emotions, and infer what they may be feeling are themselves important aspects of cognitive function, usually classified as 'social cognition'. There is now quite a bit of evidence that social cognition, and the brain areas that it depends on, are changed substantially by pregnancy. You can think of this as being a sort of cognitive reorganisation occurring in preparation for the mother's new role: making more resources available for care-giving skills, at the expense of other aspects of cognitive function that will be less important to her offspring's health.

In December 2016, researchers in Barcelona reported the first evidence that this cognitive reorganisation can be seen in the brain. They did this by taking brain scans of first-time mothers before and after they became pregnant. When they compared the scans, they saw significant reductions in grey matter in a network of regions that are associated with social cognition. The researchers then went one step further, following up the women two years after the birth. The brain changes they had recorded were still present, and the amount of change seen was related to scores on a test of maternal attachment. So the most-attached mothers were those who showed the biggest brain adaptations during pregnancy, suggesting that this is a mechanism by which the brain fine-tunes itself to meet the challenges of child-rearing. The changes were so clear that a computer algorithm could correctly predict whether or not a woman had been pregnant based on just her brain scans.

So, big biological cycles can have big effects on our brains. For some people, employment drives cycles in biological functions that are likely to impact their mental well-being and other aspects of brain function. Think of the monthly pattern of stress levels, for example, in a salesman with ambitious end-of-month targets, or the disruption of circadian rhythm experienced by airline crew dealing with constant jet-lag, or workers with rotating night-shift patterns. Indeed, even for people with nine-to-five office jobs, patterns of behaviour may differ quite widely between working and non-working days. But the biggest and most obvious constituent of our non-working time, around a quarter of a million hours across a lifetime, is sleep. And it turns out that sleep is really, really important in brain function.

The magic powers of sleep

Sleep does an awful lot of useful things to our brains. For example, sleep fixes new learnings more permanently into our memory; to use the technical term, it helps with memory consolidation. Many experimental studies have shown that sleeping after you learn new facts, experience new events or practise a new skill helps consolidate that memory. This is not only true for a full night's sleep but also for much shorter durations: in some studies, naps of just a few minutes have been shown to help with memory retention, and the effects can be detectable as an increased likelihood of recalling an event many years later.

You probably know that during a whole night's sleep, you move through different stages of sleep, in cycles of about ninety minutes. The amount of time you spend in the different stages changes not only as you get older, but also throughout the night. EEG studies show that at the start of a good night's sleep, we tend to spend more time in slow-wave sleep (SWS), which is often referred to as 'deep sleep', whilst in the latter part of the night we spend more time in rapid eye movement (REM) sleep.

As a result, even subtle changes in sleep pattern – waking up earlier on a Monday morning than you did on Sunday, say – can result in different doses of the different stages of sleep. This is important because the processes occurring in the brain during the different sleep stages seem to do slightly different jobs with respect to memory. In particular, SWS seems to help with the consolidation of declarative memories, things that you can consciously recall. In contrast, REM sleep seems more important for emotional memories and procedural learning.

As we get older, the time we spend asleep tends to become

shorter and more fragmented, leading some to speculate that it might explain some of the reduction in cognitive functions that happen in both healthy ageing and neurodegenerative disorders such as Alzheimer's disease. At the other end of life, teenagers are notorious for having sleep patterns that don't necessarily fit well with educational and family schedules. If this mismatch between teenagers' circadian rhythms and the demands placed on them by society lead to sleep deprivation, this could be a problem. Experimental studies in small numbers of teenagers show that sleep deprivation tends to worsen performance on tasks that require them to maintain a high level of vigilance. Conversely, measures that increase the amount or quality of sleep tend to provide a boost to working memory. So sleep-deprived teenagers would clearly not be in the best cognitive state to benefit from their time at school. In fact, sleep deprivation produces cognitive impairments in all ages: in one brave study, toddlers were experimentally deprived of their usual daytime nap and then challenged to solve an unsolvable puzzle. These nap-deprived two-to-three-year-olds behaved like younger children, tackling the task in a less effective manner and being far more likely to persevere even when faced with evidence that the puzzle couldn't be completed.

Sleep is so important that many of the things that are popularly thought to affect our cognitive abilities – such as being ill, hungover or jet-lagged – are probably just reflecting the cognitive effects of disturbed sleep patterns. Interestingly, when sleep disturbance happens to the whole population, as it does twice a year in countries that observe daylight-saving time, these small differences in alertness and decision-making power can be seen in the number of fatal accidents that occur:

significantly more the morning after the clocks 'spring forward' and many people lose an hour in bed, and fewer the day after they 'fall back' and people get an hour's more sleep. On that basis, we might be wise to take extra special care of ourselves and our actions at the time of the spring change.

Optimising your day

Other things that vary between one day and another have relatively small influences, though these may have an important cumulative effect over time. For example, we know that the brain needs adequate nutrition to function, and that long-term nutritional deficiencies can cause all sorts of nasty brain problems. Vitamin B12 is needed to maintain good myelin (the fatty sheaths that cover neurons so they can send long-distance signals efficiently), and both vitamins D and B9 (folate) have been linked to cognitive function and may be important in various forms of neurological and psychiatric illnesses. Shorter-term deprivation, such as a period of low blood sugar, can also make quite a difference to the brain's ability to function, and at the extreme of hypoglycaemia (usually experienced by diabetics or endurance-sports fiends) it can completely wipe out complex cognitive functions. At less extreme levels, it's hard to know just how much difference something like having a good breakfast before an exam makes. A lot of research into this has been carried out by breakfast cereal companies, which, as you would expect, have a lot at stake in investigating this. Therefore much of the research is not entirely bias-free. None the less it's fair to say that, on the whole, having a breakfast that provides slow-release energy for the brain is a good start to the day, whether measured in terms of the quality of children's schoolwork or formal testing in adults.

What of Sally's pre-breakfast run? There is a very strong link between cardiovascular health and brain health, and it's clear that people who take regular exercise have a lower risk of brain disorders such as stroke, dementia and depression. But it's hard to motivate yourself to put on your running shoes based on a tiny change in risk of developing a disease several decades in the future. What would (perhaps) be more motivating is to know more about an instant improvement. In other words, if you have a big presentation to give today, should you lie in bed for thirty minutes more, or go for a run? Assuming you have already had a good night's sleep (not always true the night before a big day), the answer would likely be: go for a run. There is now considerable evidence that physical exercise can produce acute short-term improvements in cognitive function, as well as affording long-term brain health benefits, which we will discuss more in Chapter 8. But for now, just accept that really you have no excuse . . .

Moment by moment

How much the brain fluctuates moment to moment is not a topic that receives a lot of attention. If anything, researchers typically consider it an annoyance that measurements of cognitive performance, regional blood flow or any other brain markers might be affected by the time of day, day of the week, or when in the year they are running the experiment. These things, along with what the participant had to eat last night, whether they usually have coffee in the morning and whether or not they smoke, are usually taken to be just sources of random error which, all being equal, should even out across the whole experiment and thus not affect the results too much.

But even such small effects can have real consequences. One of the most startling research studies of recent years is an analysis of the effect of time of day on decisions made by some experienced Israeli professionals. Like the rest of us, the participants' decision-making abilities seemed to be considerably affected by tiredness, and by coffee and lunch breaks. Unlike most of us, however, the decisions in question had an immediate impact on people's lives, because these professionals were judges working on a parole board. The size of the effect was astonishing: the first prisoner who came up for parole each day had about a 65 per cent chance of being freed, but for the last prisoner seen before a refreshment break that chance was close to zero.

Safety-critical professions, such as those that involve operating large vehicles and machines, have rest breaks and time-limited shifts enforced by law or employers. For the rest of us, the consequences of operating at suboptimal cognitive capacity are perceived as much less dangerous. You can bet the Israeli prisoners denied parole wouldn't feel that way.

One other big driver of moment-by-moment cognitive performance is harder to measure but enormously important: how much pressure you are under, and whether that has built up over time to produce stress. Although it's not a specific medical diagnosis, we all know what stress means, and what it feels like. And unless you're reading this from a parallel universe of extreme serenity, you will probably have experienced the physical effects of the acute 'fight-or-flight' response: a chemical cascade of adrenalin and other hormones that is controlled by the sympathetic nervous system. Its goal is to release energy and prepare the body for physical exertion; the churning stomach and dry mouth are unfortunate side-effects of these changes. This physiological arousal would be very useful if you needed to run away from a tiger. However, it's not exactly what you want when you're about to walk into a job interview, or give an important business presentation or a big speech. What you really want there is the ability to think clearly, speak steadily and deal calmly with any difficult questions that arise.

So what effect do the acute changes in neural chemistry caused by the fight-or-flight response have on psychological state and cognitive function? Anecdotal experience suggests that it can go two ways. Sometimes the added pressure gives extra clarity of thought, like a singer giving the performance of his or her life when in front of a huge audience. Other times the pressure is too great, and we clam up. The idea that physiological arousal improves performance up to a point, after which it worsens it, is known as the Yerkes-Dodson law. It was originally derived from experiments that studied the effects of electric shocks of varying intensities on learning in mice, and despite a lot of data challenging it, it remains a popular theory that crops up in most introductory psychology textbooks.

It's a long way from mice receiving electric shocks to the kind of acute and chronic stressors that are commonly experienced by affluent modern humans. So attempts to understand the relationship between arousal and performance have started to get a little closer to home. In one recent study, ninety-one people were randomised to prepare and deliver a speech in either a high- or low-stress situation. The high-stress participants were told to defend themselves against a (fictional) shoplifting charge, while those in the low-stress group made a short video summarising a travel article they had just read. The researchers carefully coded every utterance in the speeches, and also measured heart rate and levels of the hormone cortisol, which is released in response to stress. As you would expect, people asked to defend themselves against the shoplifting charge showed higher cortisol levels and higher heart rates while speaking. But the quality of speech didn't seem affected by this stressor. The two groups used the same number of words, on average, and those in the low-stress group actually used 'non-fluencies' like 'umm' and 'er' more frequently. Interestingly, people in the higher-stress situation tended to pause more during speech, which the researchers suggest was because simultaneously dealing with the stress was reducing the amount of cognitive resources available for giving the speech, making them slower to form their next idea and phrase.

We hope you're so absorbed in reading this book that you're largely unaware of what's going on around you – the conversation between the folk across the aisle, the kids playing outside, or even just the minor changes in the noise, light and temperature of your environment. If you were to register all of these things equally and simultaneously it would be extremely hard to concentrate on anything at all, and you almost certainly wouldn't be able to read, understand or remember any of this paragraph.

Given the large amount of sensory information available to us at any one moment, we are extremely good at directing our attention to a very small subset of it. There are whole shelves of psychology textbooks devoted to discussing how exactly filtering out of irrelevant information takes place, and under what conditions we are most able or unable to do so. Is our attention like an adjustable spotlight, such that we can take in a wide array of information at a shallow level, or alternatively a narrow array of information in great depth? When we think we're multi-tasking, are we actually just switching rapidly between two different demands on our attention? Can we process different types of information in parallel, and what limits our ability to do so? Aside from keeping psychology and neuroscience teachers in business, these questions matter because they define, at a functional level, our neural limits. One very interesting limit, which reflects brain function at the millisecond level, is temporal: how long is attention captured for by any one stimulus?

Imagine you're playing a computer game where a bunch of ravenous aliens fly rapidly towards you on the screen. Your job is to shoot the bad aliens, and not shoot the good ones. Here comes a baddie: bang – got him! But you didn't even see the second baddie who showed up immediately after the first. And he ate you. What killed you there was the 'attentional blink', one of the most researched phenomena in cognitive psychology. The attentional blink is interesting because it tells us about how rapidly the brain can shift attention between individual packets of information. In general, we are good at processing incoming visual stimuli very rapidly, but if one of them is a target for our attention, we then show a strange form of blindness. In our computer game, a second incoming target that occurred immediately after the first, say within a fifth of a second, would likely be noticed. But then there is a period of around a quarter of a second in which incoming targets won't make it to your conscious awareness: you would see, but not notice, one if it appeared during that timeframe. Some people think this temporary attentional blindness reflects a limit in the capacity of the brain to process information during this time; others believe it's a mechanism for preventing extraneous information leaking into higher states of processing, such as working memory, alongside the intended target.

How much brain are you using right now?

We started this chapter by noting that the brain changes considerably over the timespan of years: developing, peaking and then declining in function over a lifetime. What we've learned from our study of navigating birds, pregnant ladies and parole-board judges is that subtle changes in what the brain can do, and how it does it, can change over much shorter time-frames too. The brain is very adept at supplying the functions you need moment by moment and, over time, differences in how you spend your moments accumulate into longer-lasting and bigger changes in the brain, as we saw in the London taxi drivers. How much brain you need right now depends of course on what you're trying to do – but also on what you were doing yesterday, last week, last month, and just milliseconds ago.

So there is great variation in function within the healthy brain. Part Three of this book takes us beyond normal variation and considers how the brain responds to extraordinary circumstances.

A VIEW FROM

Dr Simon Kyle, Senior Research Fellow at the Sleep and Circadian Neuroscience Institute, Nuffield Department of Clinical Neurosciences, University of Oxford, UK

Simon is a sleep researcher at an exciting new multi-disciplinary research institute in Oxford that aims to understand the relationships between sleep, circadian rhythms and health. He is particularly interested in what causes sleep disturbance and how it can best be managed, and also researches the interactions between sleep disturbance and other aspects of mental health. Simon is the course director for the Oxford Online Programme in Sleep Medicine, an innovative postgraduate course that is training the sleep specialists of the future.

Jenny speaks to Simon, ironically, after a really terrible night's sleep. So she's keen to hear what he can tell her about why we crave sleep so much, and what effect a lack of sleep last night might be having on her brain today.

Simon, can you give me an overview of why we need sleep?

'Well, in biology, to understand the function of something we often disrupt it. Most of what we know about the effects of sleep on the brain comes from two different ways of looking at sleep deprivation. In one, we run very controlled experiments where we take people who have a normal pattern of sleeping, and then either restrict their sleep by a number of hours or totally deprive them of sleep. When we do this we find that it reliably worsens cognitive performance, on measures like vigilance, sustained attention and working memory. Restricted sleep down to five hours a night, over several nights, has a cumulative effect on cognition, leading to increasing impairment with each day. Interestingly, some studies have shown that a couple of nights of catch-up sleep – like we might do on the weekend – does not fully restore this performance impairment. Sleep plays a central role in the consolidation of memories. Studies clearly show that this sleep-dependent memory benefit is impaired when you experimentally disturb sleep, and our ability to learn information the next day is markedly reduced after sleep deprivation.

'But we can only sleep-deprive people for a few days at a time, for ethical reasons. So the other way we look at the effects of sleep loss is by studying people who report long-term problems with sleeping. We define insomnia as having difficulty either initiating or maintaining sleep, and in large studies such as the UK Biobank we find that around a third of people report experiencing frequent insomnia symptoms in the last month. For about 10 per cent of people, these problems last for more than three months and affect their daytime function. These people have a higher risk of going on to develop a whole range of mental and physical health problems: depression, anxiety, substance abuse, heart disease, stroke, Alzheimer's disease. We don't yet know for sure whether the sleep problem causes the other disorder, or whether it is an early symptom of it. But we know from neuroimaging studies that people who have insomnia but no other symptoms of disease do show brain changes, such as shrinking of the cortex, suggesting that chronic sleep disturbance may be having a negative effect on the brain.

'Interestingly, there are some people who seem very resilient to sleep deprivation. So while on average we see a significant worsening of cognition, there is a small minority of people who seem to function just as well after a few nights of sleep deprivation as they did before. And there are other people who seem particularly vulnerable to the cognitive effects of sleep loss. There are some interesting individual differences and we don't yet understand what causes them.'

How much should people worry if they feel they aren't getting enough sleep?

'I think the first thing to say is, most people probably are getting enough. Even though the media talk a lot about the effect of technology and these very busy lives we all lead, meta-analyses show that the amount of sleep we get, on average, hasn't changed over the past 50 years. Lots of people restrict their sleep during the working week, choosing to work long hours or socialise instead of prioritising sleep, but then make up for it at weekends. The number of people who report having insomnia has gone up over time. But for about a third of people who report having insomnia, when they come into a sleep lab, we find that there is a sizeable discrepancy between how much sleep they think they get and how much time our instruments show they are asleep. So for example someone might think they are only getting five hours' sleep, but when we measure it they are getting seven hours.'

So when should people worry?

'It would be a real concern if you were so sleep-deprived that you were at risk of having an accident, or if your work was suffering, for example. And people underestimate how long it takes to bounce back from a period of sleep deprivation. If we bring people into a research study and restrict their sleep for five nights, and then let them sleep for up to ten hours a night for two nights, their cognitive function still won't be back to normal at the end of it.

'But for many people with sleep problems there are relatively simple things that can be done to help, such as taking regular exercise, developing a regular pattern of light exposure and sleep schedule, and avoiding daytime naps and stimulants. If you are doing all these things, and giving yourself enough time to sleep but not getting as much as you need, or experiencing persistent poor-quality sleep, then your GP can help. For those with long-term sleep problems cognitive behavioural therapy [CBT] is the treatment with the highest level of evidence. CBT is a structured psychological therapy which tackles thoughts and behaviours that maintain poor sleep. Often the key is to create the right conditions pre-sleep so that we can de-arouse mentally, enabling our biological drive for sleep (sleep pressure and circadian rhythm) to overtake wakefulness.

'As we learn more about the neuroscience of sleep we are realising that sleep probably isn't an all-or-nothing thing. Recent studies here in Oxford have shown that if you record from individual neurons in mice, you will find that within the same brain, some groups of neurons will be in a sleep state and others in an active state, all at the same time. We call this "local sleep". So sometimes when people have an experience of being aware of things during the night, the brain may overall have been asleep at that time, but with some populations of neurons awake, thinking and processing information. This may be one reason why people feel they are getting a worse night's sleep than they actually are.'

How much brain do we really need?

'I think we need all of it! Though perhaps we need a more personalised view of the brain – the extent to which we engage different brain regions for different tasks varies from individual to individual. We are also getting a better sense, from work on unconsciousness and coma, that there can be spared cognitive abilities even in these states.'

PART THREE

Beyond the Limits

How much brain can we afford to lose?

CHAPTER 6

Something Missing: Can we function normally without a complete brain?

Let's think now about how the limits of human brain capacity are stretched when it suffers abnormal challenges. You may have heard tales of people finding out that they've lived for years with a bit of their brain missing when they go for a scan for some other reason. We're led to believe that such individuals were carrying on completely normal lives with no apparent signs that anything was wrong. Such urban legends suggest that the brain is highly adaptable even when significant chunks are missing. But is this really true? Does this indicate that we can, in fact, manage just fine without some of our brain and, if so, how much can we potentially afford to lose?

Here we take a look at cases of people who have experienced life with part of their brain missing from birth, and others whose brain was complete one minute and the next not, whether as a result of trauma or surgical intervention. We consider how they have fared in life and what this might tell us about brain adaptability (known as plasticity, or 'neuroplasticity', to be exact) and essential functions. We also have a think about what might be happening to allow people to function without an anatomically complete brain. Do other parts of the brain somehow compensate for the missing pieces? And does timing play a role? If a part of the brain is missing from birth, is the rest of the brain better able to take over the missing functions as it grows and develops, or is a fully grown brain more robust against trauma?

One way to reflect on how the brain copes is to take an anatomical tour, looking at some of its different components in turn. In this way we can think about what those parts are responsible for and how a person might be affected if a bit is missing. We will then find out what happens when bigger parts and more areas of the brain are affected.

Starting from the top: the cerebrum

As we learned previously, the cerebrum is the largest part of the human brain. As a whole it is responsible for myriad functions, including higher functions such as emotions, learning and reasoning, and interpreting touch, vision, hearing and language. The cerebrum is divided into two halves, called hemispheres, and each of these is comprised of four main lobes: the frontal, parietal, temporal and occipital lobes. Each of these has a range of different jobs.

The case of the missing parietal lobe

Cole Cohen is a published author. She has a Master in Fine Arts degree in writing and critical studies from the California

Institute of the Arts. She was a finalist for the Bakeless Prize and the Association of Writers & Writing Programs Prize in Nonfiction. At the time of writing she lives in Santa Barbara, where she has a job as the events and programme coordinator at the University of California. Yet, despite her above-average intelligence, she has experienced learning difficulties throughout her life that have made many seemingly mundane tasks impossible.

Cole struggles to judge time and space. She can't gauge the passing of time without the aid of a watch; for instance, she could stand by the side of the road and be unable to estimate whether an oncoming car would arrive in ten or thirty seconds. She wouldn't know whether one minute, ten minutes or an hour had passed. She wouldn't know how long to hug someone. She gets lost in the supermarket or even on the way to a familiar destination. She cannot fathom numbers and money is a mystery. Yet, despite having such problems and being examined and tested by numerous professionals, it took many years to discover what was wrong. Cole was twenty-six years old when she found out that she had a hole in her brain the size of a lemon. This hole is in her left parietal lobe, which, you won't be surprised to hear, is responsible for spatial awareness, perception of objects and mathematical abilities.

Cole's response to the discovery of the hole was to wonder why she wasn't dead, or at least highly cognitively damaged. Her doctor explained that it is because the hole is in the parietal lobe and not the frontal lobe that she is still very well and highly functioning. So it seems that having a complete parietal lobe is not essential to human survival. At the very least, if you're missing a considerable chunk of it from birth, you can still develop well and get on with life. Cole has even written and published a book about her brain (*Head Case: My Brain and Other Wonders*). But her case is very, very rare. In fact, as far as she is aware there hasn't been another recorded case like hers. There may of course be others who have a hole in their parietal lobe but these just haven't been detected or written about yet. If there are others out there with a similar missing piece, we don't know if they are also doing well or are suffering many problems as a result.

What about the frontal lobe then? Cole's doctor suggested that if the hole had been in that part of her brain instead, the picture would have been more bleak.

The case of the missing frontal lobe

The frontal lobe is the largest lobe in the brain and sits, unsurprisingly, at the front of the cerebrum. It is involved in many functions, but particularly those that make us who we are as individuals. These include emotional expression, social and sexual behaviour, judgement and impulse control, spontaneity, language, memory and motor control. When something goes wrong in this part of the brain it can affect our very being; our personality. If this happens, it can go on to change our existence in the world and our relationships, as we are no longer the people we once were.

Significant changes in personality following trauma to the frontal lobe have been linked to dramatic criminal acts. An accident at work that happened in a split second changed the personality of family man Cecil Clayton back in 1972. He was working at a sawmill in Missouri when a piece of wood flew up into his head. It pierced his skull and brain, leading doctors to remove almost a fifth of his frontal lobe. He went from being a religious, sober, hard-working, happily married father to suffering uncontrolled rage, hallucinations, confusion, paranoia and suicidal thoughts. In 1996 he was convicted of murdering a police officer but made news headlines in 2015 when he was scheduled to be put to death for this crime. It was argued that he was mentally unfit for capital punishment due to the accident that changed his personality, and that the Supreme Court had since banned judicial killings of insane and intellectually disabled people. His execution proceeded nonetheless.

Elsewhere in the US, Kevin Wayne Dunlap was sent to Death Row after he admitted to killing three children and attacking a woman in her home in Kentucky. His lawyers described his behaviour as perplexing. They said that he was impulsive and unable to behave rationally, both at the time of the crime, making no effort to hide his identity, and also when he pleaded guilty out of the blue although it was uncertain that he even knew what he was admitting to. Six days before his trial it was discovered that a large part of Kevin's frontal lobe was either damaged or missing, although this information was not considered relevant by prosecutors.

In both these cases it is not known whether the damage to the frontal lobes was in any way responsible for the men's criminal actions, as most people who suffer frontal lobe damage do not go on to commit crimes. However, the cases do provide a small glimpse into how physical changes in the frontal lobe can seemingly lead to big changes in a person's behaviour.

A well-known name in the annals of psychology is Phineas Gage. In 1848 he suffered the unfortunate accident of a small crowbar passing completely through his head. At twenty-five years old he was the foreman of a railway gang in Vermont, and a hands-on kind of guy. He was involved in the preparations for rock-blasting with dynamite when he accidentally triggered a blast too soon, causing the crowbar-like tool, a tamping-iron, to fly into his face under his left cheekbone, through the back of his eye and out of the top of his skull. Astonishingly, Phineas survived this injury, despite being in the countryside, away from a major city with medical expertise, and having only the medical support that the nineteeth century could offer. It was more than an hour before he saw a doctor, who stemmed the bleeding and managed to prevent him succumbing to infection. It took some time to recuperate and he couldn't return to his old job, but eventually he became a stagecoach driver and then a farmer. Despite such a serious injury, Phineas lived a productive life for many years. However, although he seemingly made a good recovery without any paralysis of the body, for example, he didn't make a complete one: reports suggest he underwent a significant change in his personality. His former railroad employers described how he used to be regarded as efficient and a 'smart, shrewd businessman' but following the accident became irreverent, impatient and indulging in 'the grossest profanity', with a childlike intellectual capacity. It is estimated that the tamping-iron passed through his prefrontal cortex. Research has shown that injury to this region of the brain can cause profound personality changes, without other apparent neurological problems occurring. The prefrontal cortex is linked with memory, personality and the ability to moderate one's behaviour, which helps to explain the changes observed in Phineas.

British Olympic Gold medallist rower James Cracknel OBE also famously suffered a significant head injury in 2010, when he was struck by a petrol tanker while cycling in the US. He fractured his skull and suffered bruising to the brain. According to his doctors, during the accident his brain had swung forward, hitting the inside of his skull, and the impact damaged the frontal lobe. This injury left James with memory loss and he became easily frustrated, temperamental, stubborn and impatient. In addition, he can no longer smell or taste.

The frontal lobe is so complex, and responsible for high-level, sophisticated functions, that it takes a long time to mature, continuing to develop into our twenties. Does this mean there is a longer window of opportunity for it to repair and rewire if something goes wrong? Or does it mean that, as such complexity requires much longer to construct, if anything happens during this time we may never reach the level of sophistication typically achieved in adulthood? We'll consider the effect of damage to the brain during childhood a little more towards the end of this chapter.

The trouble began in the temporal lobe

And now for something completely different: an uninvited guest. A fifty-year-old man in the east of England had the misfortune of being the not-so-proud owner of a tapeworm that burrowed its way across his brain. When he took a trip back to his homeland, China, the tapeworm jumped aboard and stowed away inside his brain for a free ride across the world. The unsuspecting man subsequently suffered a range of symptoms, including headaches, seizures, episodes of altered smell, memory impairment, and pain on his right side. An

MRI scan revealed areas of damage (typically called 'lesions' by neurologists) in his right temporal lobe. Initially, tuberculosis was suspected to be the problem and the tapeworm went undetected. A series of scans was taken over four years, which showed the brain lesions to be moving from the right to the left hemisphere; in other words, they revealed the path of the worm. Eventually the worm was identified via a biopsy and removed. The man was given a drug against parasitic worm infestations and was described in 2014 as being well.

The mere thought of something eating through the brain makes one shiver, but what makes it worse in this case is that the tapeworm was in the man's brain for years and its journey was being captured on film. Doctors on the case said that the various neurological symptoms the patient suffered changed in nature over the course of the infection. This is presumably because as the tapeworm travelled across the brain it affected different parts, each of which has different functions. Although evidence of the tapeworm was initially seen in the temporal lobe, as it moved it damaged more of the brain, causing a wider range of symptoms.

A brain of two halves

As the case of the tapeworm invasion highlights, people may be affected in more than one area of the brain and so experience a host of effects. However, the cerebrum doesn't work just as four discrete lobes. It also coordinates functions, as a whole and in halves (hemispheres). Normally, the left side of the brain controls the right side of the body and the right side of the brain controls the left side of the body. The two hemispheres are not mirror images, however, as they have particular functions of their own. The left hemisphere is more involved in language and communication, and detailed analysis of information, whereas the right hemisphere has an emphasis on spatial awareness, interpreting and remembering visual information, and combining information to create the big picture, for example. But for some people, what the two sides specialise in is not the most important thing; just having one at all is a relief. These people are not blessed with two fully intact hemispheres. They may be missing a part of, or indeed a whole, hemisphere.

The case of the missing hemisphere

In Germany there lives a young lady who was born with only one hemisphere of the brain. She was sent for a scan when she was three and a half years old after experiencing brief involuntary twitching on her left side; it showed that her right hemisphere was missing. While she had some weakness on one side of her body, her twitching was successfully treated and she was otherwise well. She went on to go to school and was able to master activities that require coordination of both sides of the body, such as rollerskating and riding a bike. Lars Muckli, a neuropsychologist from the University of Glasgow who led a study into the girl's vision when she was ten years old, described her as 'witty, charming and intelligent' and said that she has normal psychological function and is able to live a fulfilling life.

The German girl also only has one functioning eye: the left one. However, Muckli was particularly interested in her vision as it is very special – so special that it can do something that no one else has ever been found to do. Usually, our brain receives different visual messages from each eye and puts these together to create a single, complete picture of what we are seeing.

In essence, we use both our eyes together to form a whole image; this is known as 'binocular vision'. However, the German girl has this complete vision from just one eye. Normally, the information received by each eye is mapped on the opposite side of the brain; i.e. information from the right eye is mapped on the left hemisphere and vice versa. What is fascinating about the brain of the girl from Germany is that nerve fibres from her functioning left eye that should have linked to her missing right hemisphere had diverted to the left hemisphere. In addition, areas of the left hemisphere had adapted in order to process the left visual field, which is normally the job of the right hemisphere. This rewiring in the brain means that this girl has near-perfect vision from one eye.

It is quite remarkable what the brain can do to remodel itself in the face of adversity. In response to the findings of his study, Muckli remarked that whilst we know that the brain can demonstrate amazing plasticity, it was astonishing just how well the remaining hemisphere in the young lady's brain managed to adapt to compensate for the missing half. Whilst there are no other recorded cases like this, there may well be others out there whose brains have adapted in a similar fashion.

Opposite to the girl in Germany is Michelle Mack, who is missing her left hemisphere. Although her parents knew something wasn't quite right when she was just a baby, it wasn't until she was twenty-seven years old that a brain scan revealed the reason. Whilst the left side of the brain is typically associated with language and communication, in Michelle's case it seems that her remaining right hemisphere has taken over some of these functions that would have been otherwise lost. In fact, she has fairly normal language abilities – but it seems this may have come at a cost. The right side of the brain is typically involved in visual and spatial processing, and these are troublesome for Michelle. Dr Jordan Grafman, chief of the Cognitive Neuroscience Section at the National Institutes of Health, USA, who diagnosed Michelle's problem, speculates that during her development she failed to acquire all the skills associated with the right side of the brain because it was too busy taking on some of the abilities normally covered by the left side. Although Michelle experiences a range of problems, from difficulty controlling her emotions to becoming easily lost, it is impressive how much she can do when such a large part of her brain is missing.

Of course, these cases raise the question of whether this rewiring has only been possible because a hemisphere was missing from birth, and as the brain grew it was able to develop and adapt. But how adaptable might the brain be if you lost a hemisphere when older, once your brain had already developed to some extent?

When a whole half is removed deliberately

Imagine being told that surgeons need to remove half of your brain. Now this isn't likely to happen if you have a healthy, normally functioning brain, yet some people who experience severe seizures that cannot be controlled with drugs do face this mind-boggling decision. Seizures are the result of unregulated electrical activity in the brain. This activity typically starts out in a discrete area of the brain and can spread to other regions. However, in some people the activity may have no original focal point and spring up in multiple places within one hemisphere.

Removing one of the brain's hemispheres stops the electrical impulses travelling from one side of the brain to the other, and so reduces seizures.

For some patients a hemispherectomy may involve removal of just a part of the hemisphere, but for others the whole half is removed. It sounds pretty extreme and something that might have been the practice of Victorian physicians, who were not known for subtlety in their methods. It's true that the hemispherectomy has been around for a long time, but it has been refined over the years since it was first demonstrated to be an effective treatment. The first known example was carried out on a dog in the late nineteenth century, but the first to take place on a human was in the 1920s, at Johns Hopkins University in Baltimore. Although you might imagine this to be a very rare procedure, over 100 hemispherectomies are carried out each year in the United States alone, and it is recorded to have excellent outcomes, in the grand scheme of things.

Hemispherectomy can take place at any age, but younger patients seem to do better as the remaining half of the cerebrum often takes over the functions of the hemisphere that has been removed. Studies of children who have undergone hemispherectomies have found that not only are seizures reduced but sensory, motor and language functions are recovered as the other side of the brain develops new capabilities. It should be noted that the severe seizures themselves can inhibit normal development, so removal of the troublesome hemisphere can open up the brain's opportunity to progress. The plasticity of the brain is never more clearly demonstrated than in these cases, where such a large proportion is removed and yet significant rewiring enables the brain to adapt very effectively.

Take the case of Aiden Gallagher. In 2003 he featured on *NBC News* as an ordinary kid with an extraordinary background. He was a ten-year-old from Ohio, going to school and enjoying playing sports, but since the age of three had been living with only half his brain. As a toddler severe seizures were having a huge impact on his life and his development. His father explained how Aiden had forgotten his alphabet and how to count and that his previous knowledge seemed to be slipping.

Aiden had a hemispherectomy, with a good recovery after the surgery, running around a local playground within a week and not having a seizure since the operation. He was just one of 186 hemispherectomy patients whose cases were reviewed in a study by US researchers Ahsan Moosa and colleagues to understand more about their long-term outcomes. They found that the procedure was successful in significantly reducing seizures and patients did well on the whole, although many did experience impairments in reading and language. It's worth noting, however, that those who undergo hemispherectomy typically suffer some loss of movement in the limbs on the side of the body opposite to the hemisphere that is removed. Vision problems on this opposite side are also common.

The case of the missing corpus callosum

Just as the four lobes of the brain don't work in isolation, neither do the two hemispheres. They are joined by something called the corpus callosum, a bundle of neural fibres that transfers motor, sensory and cognitive information between the hemispheres.

Not everyone possesses a corpus callosum. Although it is a rare condition, agenesis of the corpus callosum, where there is a partial or complete absence of this structure, is among the most common brain malformations, with between 0.5 and 70 cases per 10,000 population. As you can imagine, this can lead to communication problems in the brain. Surprisingly, however, the effect of missing a corpus callosum appears to differ between those born without one and those who have had it surgically removed later in life. Those who have their corpus callosum surgically removed typically experience a failure to transfer information between the hemispheres, known as 'disconnection syndrome'. In contrast, those who have never had one don't appear to have this syndrome, and communication between the hemispheres happily continues. This mystery flummoxed neuroscientists for decades. In recent years, a team of researchers from Rio de Janeiro and Oxford used brain-imaging and psychological tests to take a closer look at people born without a corpus callosum. Their findings suggest that in these people the brain extensively rewires itself, generating new circuits to compensate for the lack of the normal route of communication. The researchers propose that this effect can only come about at a very early stage of a person's development, when growing axons can be diverted into new pathways. This, they say, would help to explain why those whose corpus callosum is removed surgically are unable to recapture communication between the hemispheres; in short, it is just too late in the day for that to happen.

Despite this amazing plasticity of the brain, things are not quite as rosy as they first appear. Many of those born without a corpus callosum do in fact suffer from numerous health problems and often have other medical syndromes. Common problems include intellectual disability, sight and speech problems, seizures, feeding problems and behavioural problems.

However, the extent to which individuals are affected by such problems ranges widely from 'subtle' to 'severe'. In practical terms, those missing a corpus callosum typically experience delays in attaining developmental milestones such as walking, talking or reading; poor motor coordination, particularly with skills that require left and right coordination of the hands and feet (e.g. riding a bicycle); and mental and social processing problems that become more apparent with age.

There's more to the brain than the cerebrum

That's enough of the cerebrum for now, as behind and underneath the cerebrum lies the cerebellum, or 'little brain'. Well, that's where it should be if you have one.

One day a twenty-four-year-old woman went to hospital in China complaining of dizziness, an inability to walk steadily and nausea and vomiting. Various tests were carried out, including a CT scan of her brain. It was then that a rather significant anomaly came to light: her cerebellum was missing. There was just a fluid-filled gap where it should have been. How could this lady reach the age of twenty-four without knowing that things were not quite as they should be? She was married and had a daughter and her pregnancy and delivery had been medically unremarkable. None of her parents or siblings apparently had any neurological problems. So had she led a completely normal life up until this point with no sign of anything wrong? Had the brain completely adapted to having no cerebellum, with other parts taking up its functions?

Not quite. It turns out that this lady had been dizzy for years and had never been able to walk steadily. Her mother established that her daughter wasn't able to stand on her own until she was four years old and didn't walk unassisted until she was seven. She never ran or jumped and her speech was unintelligible until she was six. She never went to school. She has a mild voice tremor, slurs her words and has slightly impaired motor skills generally.

She was diagnosed as having primary cerebellar agenesis – only the ninth case to have been reported in the medical literature. So, as you can see, missing a cerebellum from birth is a very rare thing indeed, and even rarer to find an adult functioning pretty well without one. Many of those born with primary cerebellar agenesis have significant developmental abnormalities (it is frequently associated with other defects), and their missing cerebellum may only be discovered upon autopsy. Although this lady is not functioning perfectly normally and her development was slow, she represents an extraordinary example of how brilliantly the human brain can adapt when it is growing and developing in childhood.

Someone else who is going through life without a cerebellum is Jonathan Keleher, a thirty-three-year-old man living in Boston, Massachusetts. Unlike the Chinese lady, the anomaly in Jonathan's brain was discovered when he had a brain scan aged five years old. He too was late to sit up, walk and talk, according to his mother, but health professionals didn't know what was wrong. He had speech therapy and physiotherapy and eventually was given a brain scan that revealed the problem. Dr Jeremy Schmahmann, his neurologist at Massachusetts General Hospital, remarked of the scan that there was a lot of nothingness there. Jonathan also has problems with his balance, and the way he speaks is described as 'distinctive'. Yet despite his physical challenges, he has an office job and lives independently.

He tried driving a car but his brain simply couldn't coordinate all the traffic information going on around him as well as his reflexes and movements. But it is not just his physical abilities that are impaired. He struggles with emotional complexity and found it difficult to know how to behave in social situations or how to show emotion. Whilst most of us automatically learn these aspects of life as we grow up, Jonathan has had to be specifically taught how to do such things. According to Schmahmann, he has learned these things by training other areas of the brain to do the jobs usually done by the cerebellum.

So what does the cerebellum actually do? This structure comprises around 10 per cent of the total weight of our brain, but although it is much smaller than the cerebrum, its cortex contains more neurons as it is so densely packed. The cerebellum plays a critical role in a person's movement and coordination, including intricate movements such as using both our fingers and eyes to thread a needle. It is also involved in motor learning, the process of learning movements that stay with you in life, such as walking, talking, climbing, etc. Therefore, any problems in the cerebellum of a child can have significant effects on motor development. In essence, the cerebellum plans, adjusts and executes movements of the body, the limbs and the eyes. This is sometimes referred to as 'technical intelligence', as opposed to social intelligence. Schmahmann has been studying the cerebellum for decades and believes this part of the brain to have one main purpose, which is refining clumsy actions or functions.

The cases of Jonathan and the lady in China suggest that, whilst the cerebellum clearly has important functions, it is not critical for life. People are able to live relatively well without it, although they do require the support of others to deal with a number of challenges. We clearly do better if we have a cerebellum, but perhaps we don't need absolutely all of it. If people are able to live without one at all, maybe there is an excess of brain in the cerebellum.

Let's briefly consider that possibility. If you can live without one at all, perhaps if you do have a cerebellum, it doesn't matter how big it is? Well, some research suggests that, the bigger your cerebellum, the better your fine motor skills and verbal memory are. A study of older adults also found an association between more grey matter in the cerebellum and better general cognitive ability. In neuroscience circles the cerebrum tends to get all the attention for how brilliant it is, but it turns out that the cerebellum has four times as many neurons, and researchers are realising how many important functions it is involved with. The human cerebellum expanded rapidly throughout evolution and some researchers suggest that this is an indication that technical intelligence was likely to have been at least as important as social intelligence in the advancement of the human brain. As sensory and motor control is so important in learning complex action sequences, the evolution of the cerebellum would have enabled increasing sophistication in the technical abilities of humans. In turn, this may have paved the way for additional social capabilities and interactions, including language. For example, it seems that the cerebellum helps to coordinate syllables into fast, smooth and rhythmically organised sounds.

The brain is more than the sum of its parts

We have been focusing on people who are missing specific parts of their brain, but there are of course people whose brains have been affected in more than one area.

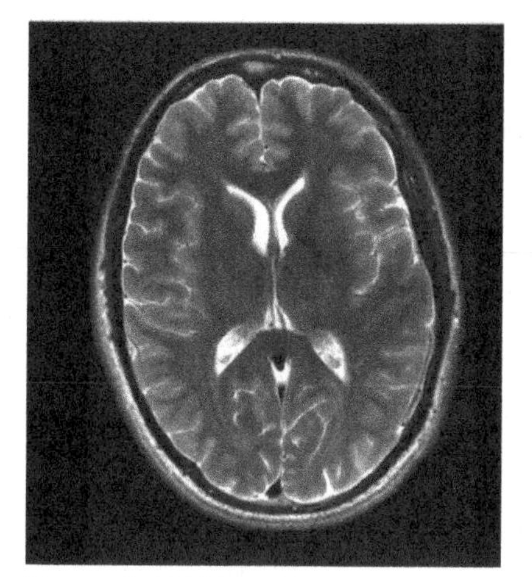

Figure 6: MRI image of a 'healthy' brain. Courtesy of The Wolfson Brain Imaging Centre, University of Cambridge

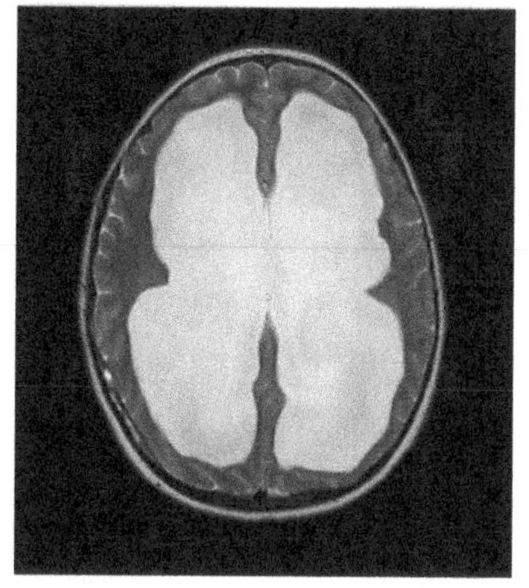

Hydrocephalus of lateral ventricle. © Living Art Enterprises/Science Photo Library

Since Sharon Parker was a child she has been told by doctors that she has no more than 15 per cent of a normal brain, and yet she now has an IQ of 113 (which is above average). As a baby she had hydrocephalus, commonly known as 'water on the brain'. An excess of fluid pushed her brain out towards the skull edge and caused her head to swell. However, by the time the problem was detected and dealt with the fluid had built up for nine months, resulting in a large hole in the middle of her brain. Amazingly, however, her brain adapted to this physical stress by reshaping itself to fit the unusual space. It formed along the skull edges, stretched and flattened its normally convoluted surface, particularly at the frontal lobes, and pushed some of its mass to the bottom rear of the skull. (Figure 6 shows an example of a brain affected by hydrocephalus versus a 'healthy' brain.) In fact, rather than just having 15 per cent of a normal brain, it seems Sharon has a complete brain after all, but in a unique form and with a big hole at its centre. Although Sharon has some trouble with her short-term memory and remembering sequences of numbers, like telephone numbers, she leads a normal life in Yorkshire. She works as a nurse and has a husband and three kids.

Sharon is not the only one living with a large hole in the centre of the brain caused by childhood hydrocephalus. A French civil servant and married father of two discovered the hole in his brain when he developed health problems in his mid-forties. Scans revealed that his brain, just like Sharon's, had been forced to the edge of the skull and in some areas the cortex had thinned in mass. He had been treated for hydrocephalus as a six-month-old. His IQ was tested and found to be below average, but he was leading a pretty ordinary life nonetheless.

The destructive path of a bullet

A US congresswoman from Arizona became international news when she was the victim of an assassination attempt in 2011. Gabrielle Giffords was attending a local political event at a grocery store when she was shot in the head. (Several others attending were also shot and six were killed.) Remarkably, Gabrielle survived this horrifying attack. To put her survival into context, it is estimated that around 90 per cent of people who are shot in the head die. Whilst there may be many factors that contributed to Gabrielle surviving, one is thought to be the location of the damage in the brain caused by the bullet. It is thought that the bullet passed through just the left side, but if it had travelled across the brain, affecting both left and right sides, she would have been more likely to die. Not only did Gabrielle survive, she has gone on to write books and co-found an organisation campaigning for better gun control in the States.

However, whilst we don't know exactly which parts of Gabrielle's brain were injured, we do know that she has subsequently experienced a range of health problems. Her right arm and leg are paralysed, her speech doesn't come easily, and she has limited vision in both eyes.

Gunshots, and in fact any object penetrating the skull and the brain, typically cause extensive damage and, in most cases, death. Whilst some individuals do survive potentially catastrophic injuries, where the injury occurs and how many critical areas it affects can make all the difference. According to the American Association of Neurological Surgeons, a bullet wound that passes through the right frontal lobe towards the forehead but well above the base of the skull is likely to cause relatively mild clinical damage because it passes through no vital brain tissue or vascular structures. However, a similar bullet passing downward from the left frontal lobe towards the temporal lobe and brain stem is likely to be devastating because it passes through critical brain tissue and is likely to injure important blood vessels inside the head.

Although some individuals do survive extreme damage to the brain from an object passing through it, it is unlikely to be a case of getting straight back to business as usual. Virtually all are left with persistent disabilities. Of course, the nature and severity of the health effects depend upon, among other things, the location of the damage. Damage to the frontal lobe, for example, may lead to quite different effects from damage to the cerebellum; the former may lead to distinct personality change, and the latter may result in clumsiness.

What if the whole brain is smaller than it should be?

In the run-up to the 2016 Olympic Games in Brazil, a viral outbreak was sweeping across South America and making its way to other parts of the world. Though causing relatively mild symptoms in healthy adults, the Zika virus was linked to developmental abnormalities in some babies born to mothers who caught the virus during pregnancy. Amongst a range of health problems in these children was microcephaly, where a baby's head is much smaller than expected. Images of babies affected in such a way were splashed across the front pages. If a baby's head is very small it is because the brain has not grown as much as it should. There are varying levels of microcephaly, which determine the extent to which the individual might be affected in life. While some children born with the condition have normal intelligence and no particular cognitive problems, others may experience any number of issues, from delays in reaching developmental milestones, hearing and vision problems, feeding problems, motor and balance problems and intellectual disability.

Another developmental disorder resulting in a brain that is significantly reduced in size is anencephaly. This rare condition is the result of a neural-tube defect, leading to babies being born without parts of the brain and skull. Sadly, those born with anencephaly die within a few days, sometimes just hours. Humans simply cannot survive this extent of developmental challenge. The limits of brain plasticity are stretched too far.

What have we learned about how much brain we need?

Humans can clearly survive, and in many cases live full lives, without a complete brain, but this typically comes at a cost. Most of those missing a part of their brain experience physical, behavioural or emotional effects as a result, which vary in their severity. However, as our case studies show, the nature of the problems depends largely on the part of the brain that is missing. Cole Cohen, missing her parietal lobe, struggles with time, spatial awareness and numbers, whereas those who have damage to their frontal lobe experience significant changes to their personalities, which change their very being in the world. Many people who have had a hemispherectomy suffer loss of movement in some limbs and vision and speech problems. These symptoms are just like those that Gabrielle Giffords, who had a bullet pass through the left side of her head, is living with. Both Jonathan Keleher and the young lady in China, whose brains are missing the cerebellum, were very late in attaining developmental milestones and are clumsy in their movements and speech. This is similar to those missing the corpus callosum, although they have particular problems with coordination, especially with things that require both sides of the brain to work together.

But what is going on inside those brains that manage to adapt to structural loss? The brain constantly reorganises itself, and brain cells are able to form new connections and re-route their activity. It can do this in response to learning and experience or to damage, whether from injury or disease. It used to be thought that the brain could only create new pathways when it was still in its development, i.e. in childhood, but it has now been observed that adult brains are also plastic in nature and can rewire in order to adapt to a changing local environment. Whilst that's the general principle, what exactly is happening at the molecular and cellular level during neuroplastic activity has yet to be uncovered by scientists.

Despite this, the principles of plasticity have been used to inform rehabilitation programmes to help individuals who've suffered brain injury to recover and adjust.

Research has found that substantial spontaneous recovery occurs in the weeks and months following brain injury. Not only that, but the way the brain responds to injury is similar to the molecular and cellular events that take place during normal brain development. Just as small children need a rich and stimulating environment to maximise their learning and development, individuals who have suffered brain injury need to be stimulated and to practise exercises over and over again, on tasks aimed at either physical or cognitive improvement, in order to grow new brain cells and enable the brain to adapt. However, it is likely that the timing of rehabilitation activities is key to their success in helping someone regain significant cognitive or physical function. It seems that various biological factors that promote the growth of new neurons and pathways are only stimulated during a relatively narrow window of time after injury, so there may be an optimal period during which rehabilitation activities are most effective. This is the main reason why doctors tend to refer to the first hours, days and weeks as being the most important in a person's recovery, and how they do during that time is often a good indicator of their prognosis in the longer term.

Youth over experience?

Many of the cases we've looked at appear to indicate that the brain is better able to rewire itself and adapt to missing pieces when it is still developing in childhood. It would also seem logical that it might be the same in the event of brain injury: that a child's brain is more plastic than an adult's and therefore has greater capacity to re-route pathways and retain function. But do children really recover better? Many studies have shown that children who experience brain injury can recover well and their brains effectively rewire themselves to make up for functions that might otherwise have been lost in the damaged or missing part of the brain. However, we also know that many children suffer permanent and disabling brain damage as a result of injury when young, so clearly the developing brain can only cope with so much change. Evidence suggests that brain injury can derail the normal path of human development. Just as in adults, after brain injury there may be clear physical consequences and changes in ability, such as problems with movement or the senses. However, unlike in adults, the full picture of how the injury has affected the child may not become apparent for many years. Whilst some may initially appear to keep pace with their peers in reaching developmental milestones, in later years a gap may open up. For example, a toddler with a brain injury may be able to achieve typical, simple activities expected for their age but as they become a teenager they may struggle with more complex functions, like planning their time, or with normal social and emotional behaviours. Once again, timing matters. The point of development at which a child suffers brain injury matters, as do other factors such as the nature and severity of the injury, their genetic make-up, their social background and home environment, and access to rehabilitation services.

Having said all this, adults do retain the ability to adapt and recover from brain injury as they age, though it must be admitted that older people are slower to rehabilitate from brain injury. Gloomily, many things decline with age and the brain's in-built plasticity is no exception.

It could be you!

The cases of people living with a part of their brain missing from birth are, of course, very rare. Only a handful of people are known to have been living without a complete brain and functioning relatively normally. There are bound to be more cases that we just don't know of, whether because such individuals live in places without easy access to brain scanners, or because no one has thought to give them a scan, or because they are muddling along with problems and assuming that's just the way life is. There may of course be a few people who are missing a bit of the brain but not experiencing any significant problems so haven't sought professional help or a diagnosis. Any one of us could be walking around with a brain that is missing a bit but not know about it. Unless you've had a brain scan and seen what's up there, how do you know it is all present and correct?

A VIEW FROM

Dr Fergus Gracey, Consultant Clinical Neuropsychologist at Cambridgeshire Community Services NHS Trust, and Senior Research Fellow at the Department of Clinical Psychology, University of East Anglia, UK

Fergus typically works with people who have frontal lobe injuries. He describes the clients he works with as those who can typically walk and talk and may seem well but actually have significant problems with other aspects of their life. As they initially recover, their physical ability improves but they often retain and develop problems with higher-level executive function. As a senior research fellow he is seeking to find out more about child and adult brain injury.

So, Fergus, why do you think some people's brains adapt well after injury and others with similar injuries do not?

'Of course, the less damage the better, but nevertheless there is individual variability, a small proportion of which is to do with individual differences in brains, but a lot of which is to do with the social and physical environment people are in.

'Many factors promote plasticity after brain injury. The environment makes a big difference. In studying rats with brain lesions, Brian Kolb demonstrated that those provided with enriched environments recovered better than those without. Educational level is associated with better recovery; the higher the level prior to injury, the better the recovery. There may also be a genetic vulnerability. For example, in mild traumatic brain injury a small percentage of people seem to go on to develop significant difficulties, possibly disproportionate to their level of injury. These people may be more vulnerable to harm due to their genetic make-up, although the evidence for this is currently limited. There is great variability in the outcome of childhood brain injury. Family and parental factors are really important; lower stress and higher parental warmth lead to better recovery. There are multiple factors that can affect recovery and, other than family ones, none of them is obvious when you work with people clinically.

'Intervention in the acute phase is crucial. For example, in the case of stroke, quick action to reduce the consequences of the initial injury, such as managing pressure in the brain after haemorrhage, helps improve recovery outcome. In terms of what you can do longer-term to help people recover, there are some practical actions that people can take. For example, you can do brain-training activities, where a person does a particular task or puzzle over and over again to promote activity and recovery in specific parts of the brain. However, evidence for the wider benefit of such activity and how you transfer it into a person's day-to-day life is currently limited. These tasks only work on one bit of the brain and not the brain as a whole, which doesn't make overall sense to a person's life. I like to use the example of how you could get a person to go to the gym and do specific exercises to work one bicep over and over again, then take them to a football pitch and expect them to be good at football. This won't work. You would be developing a great one-armed weightlifter, but nothing else.

'To aid rehabilitation you need to understand the person and how they were before their brain injury. If, for instance, a person used to use planning tools and diaries before they had an injury, they are more likely to use them in recovery. You aim to minimise their difficulties and maximise their abilities, using appropriate tools and strategies. You find out what their goals are and what they want from life.'

Can you tell me about any extraordinary examples of brain recovery from your clinical experience?

'All the cases I have worked with are remarkable in one way or another, on every level. You have to remember that many people with head trauma that we see have spent quite some time unconscious, in intensive care. They may have been in a situation where there are weeks when it is touch and go, and the nature of the injury can be so significant that you think the only outcome can be a bad one. Their recovery journey and how they go on to adapt is always in some way miraculous.'

Bearing in mind the extraordinary cases you have seen and heard of in your career, how much brain do you think humans really need?

'All of it. We don't give enough respect to our brains. The more I work with people whose brains are damaged, the more I become aware of the importance of it all. Having said that, there are folk with bits of their brain not working who achieve remarkable things. If you asked me which part of the brain would I least like to have damaged or removed, that's an easy answer: any part of the frontal lobes, especially the orbitofrontal or ventromedial areas. These play a key role in integrating emotion into decision-making, and underpin our ability to manage complex situations, including social interactions.

'Ultimately, if a bit of brain is dead then it is dead. So recovery is about how the brain reorganises itself and how a person readjusts to doing things. And it is not just about them as an individual. There is a theory known as "distributed cognition", which refers to the concept that cognition and knowledge are shared or distributed across our environment. An individual, therefore, doesn't need to know or remember everything themselves. Think about how much we rely on other people to notice, remember, etc. in supporting us to do the things that are important. So, if a function in the brain is weak, or indeed missing, we'll benefit if there are things in our environment to compensate for it. Imagine someone who has a brain injury and struggles to communicate. They may live with someone, such as a partner or parent, who knows them so well and spends so much time with them that they can understand what they are trying to communicate, even without fully functioning language. In effect they become a joint communication unit as the two work together to give the brain-injured person a voice.

'If we lose a bit of the brain, we will subsequently need an environment that supports it. This goes back to the importance of the richness of the environment. Of course, what happens in our environments has an effect back on our brains. As we think about how much brain we need, we should also be asking how much we need around us. The interplay between our brains and our environment is key. So in theory you can [afford to] have less brain if you have the right environmental tools in place.'

CHAPTER 7

Under Attack: How does the brain respond when parts of it deteriorate?

We have seen that the brain can adapt to living without all of its pieces in place, but how does it cope when it suffers from long-term assault? Even for those with an anatomically complete brain, there may come a time where things start to change. Our brains can be attacked by many factors, such as disease or environmental challenges that can significantly affect its functioning. Hundreds of millions of people around the world are affected by neurological disorders, and while some may quickly deteriorate and lose essential functions, others seem to be less affected and remain well for longer. Why is this the case? Does it depend on which part of the brain is affected, how well the person is otherwise, the age of the individual – or is it just luck? Can those with a higher IQ take more damage before it is evident in their everyday functioning? Can brain plasticity keep up if it is being attacked from many angles, or progressively for a sustained period?

We consider the extent to which the brain can cope with various forms of attack and what this may imply for both the quantity and quality of brain that is needed to function effectively. We look at a range of examples of the brain under attack, and how it responds, and have loosely grouped these into those conditions that are largely associated with motor function and those that are largely associated with cognitive function. We appreciate that this is a simplistic division and actually both motor and cognitive functions may be affected by many of the conditions we focus on, but this is a useful way to explore problems the brain may face. Despite this division, we'll note a common thread throughout: that of the shrinking brain. A common brain response to attack is to shrink, whether in part or as a whole. Whilst this is frequently observed, it is unclear what this shrinkage actually means in practical terms and whether it offers any indication that we can do without our whole brain.

Motor problems

Let us start by thinking about how the brain is attacked in such a way as to primarily affect motor functions. When disease slowly erodes a person's physical capabilities, how does the brain respond? There are many neurological conditions that can affect a person's motor control but we'll take a brief look at just two of these: multiple sclerosis and motor neurone disease.

Media personality Jack Osbourne, son of heavy metal singer Ozzy Osbourne, first realised something was wrong when he lost much of the vision in his right eye. He was diagnosed with relapsing remitting multiple sclerosis (MS) in 2012, at a time when work was going well and his wife had just had a baby.

MS is one of the most common diseases of the central nervous system: nearly two and a half million people around the world have it. Just like Jack, most people with MS are between twenty and forty years old when they are diagnosed. MS is a chronic disease that damages the nerves in the spinal cord and brain, including the optic nerves. The immune system attacks the protective myelin sheath that covers the nerves, and the damage disrupts communication between the brain and the rest of the body, resulting in a wide range of symptoms. The damaged myelin forms scar tissue (sclerosis), and people with MS develop multiple areas of scar tissue in response to the nerve damage, hence the name 'multiple sclerosis'.

People with MS typically experience one of four forms of the disease, which can range from mild to severe in their effects. Depending on where the nerve damage occurs, symptoms may include fatigue, and problems with muscle control, balance, vision or speech, mobility and cognition (such as problems with thinking, learning and planning). For Jack it was the vision problems that first indicated that something was going on; for others MS emerges in many different ways, perhaps in sudden mobility or balance problems.

Beyond the damage to individual nerves, there are also significant changes happening in the brain. The brain may shrink as nerve cells are destroyed by the disease. Brain shrinkage, or 'atrophy', is seen even in the early stages of MS, and is a well accepted imaging measure used in understanding the extent of disease in patients. The loss of brain volume has been found to be partly associated with the level of physical disability and cognitive impairment in MS. Intriguingly, grey and white matter seem to be affected differently by the disease, with grey matter showing more shrinkage, although it's not yet clear whether the most important damage is that caused to the grey or the white matter. If one is more important than the other, perhaps in future this may tell us something about how much we really need of each to function effectively, and whether we can afford to lose any grey or white matter without experiencing problems.

One has to wonder why, if we have an incurable progressive disease, we bother to measure brain volume loss at all? Well, first, brain shrinkage gives some biological clues as to what is happening in the brain during the disease process, which may in turn lead to the development of new treatments. Second, it's important to measure volume loss to judge whether a new treatment is working. Whilst we're not yet at the stage of being able to regrow affected parts of the brain, several clinical trials have demonstrated that treatments can reduce brain volume loss, and this may eventually influence clinical practice.

As you can imagine, since the nerve damage in MS can occur anywhere in the body, each person with the condition is affected differently and there is huge variability in people's experiences. Furthermore, over time the symptoms are also unpredictable; some people's symptoms worsen steadily, whilst for others they come and go. This is why you often hear about people having relapses, i.e. periods when symptoms get worse, and remissions, i.e. when symptoms improve or even disappear. What is causing such variability, even among those diagnosed with the same type of MS, is still unknown.

So it is not as simple as a disease taking hold and things then going downhill, as is clear from the different types of MS. This chapter is interested in how the brain responds to attack, and in MS it seems that it does not surrender at first asking. One feature of MS is the remarkable capacity for many patients to spontaneously recover from neurological problems attributed to inflammatory attacks on the central nervous system. For these individuals, during relapses most of the problems caused go on to resolve nearly completely over a few days to a few weeks. Investigation of brain lesions in MS reveals a coordinated molecular response involving various proteins and molecules that provide protection and encourage repair. Once we know more about what's going on, the clever restorative biological mechanisms involved could potentially be harnessed to banish inflammation and mend the brain, and may even be beneficial in treating other neurological conditions.

Remember the 'ice-bucket challenge'? People were sponsored to have a bucket of ice water thrown over their head and posted the video on social media. This challenge was associated with a campaign to raise money for, and awareness of, amyotrophic lateral sclerosis (ALS) research. ALS is a form of motor neurone disease, a progressive neurological disorder that causes significant motor problems.

Motor neurone disease (MND) progressively destroys motor neurons in the brain and spinal cord so that messages gradually stop reaching muscles, leading to weakness and wasting. This can affect a person's ability to walk, talk, eat, drink and breathe. Sadly, most cases of MND are ultimately fatal and there is currently no cure. There are actually different types of MND, which affect people in different ways (although there is some overlap), from the symptoms they experience to their average life expectancy. People with a rare form of MND called primary lateral sclerosis may actually go on to live a normal lifespan, but the more common forms of the disease, such as ALS, are significantly life-limiting, with most people living fewer than five years after their diagnosis.

Some studies have found that people with ALS experience some shrinkage in brain volume, particularly in the frontal and temporal areas. Brain shrinkage is a common theme of the conditions we look at in this chapter, but the location of the shrinkage is interesting in this case, given that ALS is thought to primarily affect motor function. Researchers suggest that this shrinkage in frontal and temporal areas may indicate that ALS is not just a motor disease, but also one that could affect cognition. Despite this shrinkage, evidence to date suggests that only a subset of people with MND experience behavioural and cognitive changes, although the cause of this variability remains unclear. Perhaps it takes a certain amount of shrinkage in particular areas to occur before symptoms are expressed. This point brings us to an example of someone with ALS who most definitely does not seem to be hampered by any significant cognitive deficits brought about by the disease.

When we think of motor neurone disease, we often think of the brilliant physicist Professor Stephen Hawking, who has defied the odds in living well beyond what was expected for his condition. He was diagnosed with ALS aged twenty-one and given just two years to live. That seemed a reasonable estimate at the time, given that today, with modern medical technologies that can assist longer living, the average life expectancy for ALS is two to five years. There has been much speculation as to why Hawking, now well into his seventies, should have lived so much longer than others with ALS. We know he is extremely clever, so does he have an extraordinary brain that is somehow better at defending itself from attack? (Later we will explore whether being clever is beneficial in any way for protecting our brain against disease.) Is Hawking in some way a superhuman who is able to resist extreme challenge from disease?

Part of the answer may lie in the fact that he was diagnosed very young. Typically people are diagnosed in their fifties, but it has been found that those who develop ALS at a young age seem to have much better survival rates, although it is not yet known why. This prompts many questions. Are those who get the disease at a young age somehow different genetically or biologically? Is the disease that develops in young people simply different from that seen in older people? Is the young brain somehow better able to fend off the disease than the older brain? Are young people just otherwise physically fitter and therefore better equipped to fight the disease for longer?

Many years ago Alexis lived opposite Professor Hawking in Cambridge, and would see him from time to time. Obviously, there's nothing remarkable in seeing a neighbour. However, one night when rounding the bend to her house she had to screech her car to a halt to avoid crashing into him as he slowly trundled his wheelchair down the road. Imagine what the headlines might have been!

The two conditions we looked at here are quite different despite both being neurodegenerative diseases primarily affecting the motor system. Whilst MS and its effects vary widely, people typically live a long time with the illness. In contrast, most of those with MND become rapidly disabled and die within a few years of the onset of symptoms. People with MS experience greater cognitive problems than those with MND, and those with MND experience greater physical impairments. Symptoms typically begin at a much younger age in MS than MND and it is more common in women, whereas more men are affected by MND. So, although motor functions may be primarily affected, there is a huge variation in response of the brain and the body when under attack from different diseases.

From the evidence surrounding motor conditions, it's clear that we can't easily pick out what parts of the brain we might be able to do without, since the brain responds in many distinctive ways depending on the challenges it faces. Perhaps looking at the response of the brain when under attack from things that cause cognitive problems may tell us more.

Cognitive problems

Let us now turn to diseases and other problems that primarily impair the cognitive ability of the brain. We all know that as we get older our cognitive functions slowly decline and we are not as sharp as we once were. For many people, however, there is more going on than just the normal age-related decline. These people suffer from dementia, a progressive syndrome with a range of symptoms including memory loss, and problems with understanding, thinking speed, judgement, perception, language and even the ability to carry out simple tasks. In 2015 it was estimated that nearly 47 million people worldwide were living with dementia, and it has now overtaken heart disease as the leading cause of death in England and Wales.

There are around 100 different types of dementia, and the second most common one is vascular dementia, which often appears as a series of mini-strokes. However, dementia is most commonly caused by Alzheimer's disease. Everyone has heard of Alzheimer's disease and many fear it. Surveys in both the UK and the US have found Alzheimer's and dementia to be the illnesses that people fear most, alongside cancer.

So what is Alzheimer's actually doing to the brain? A substantial and progressive attack is launched by this disease. There is a build-up of proteins called beta-amyloid and tau in the brain. As clumps of beta-amyloid stick together to form plaques and tau protein strands tangle and accumulate in particular brain areas, they interrupt the ability of healthy neurons to function properly. Signals between synapses are interrupted and neurons can no longer communicate with each other, and essential cell nutrients are no longer transported efficiently, with the result that the neurons eventually die. Without the smooth transmission of signals across the brain, thinking and memory become impaired and the brain's messages become lost.

As you can see, this disease means business, but it isn't necessarily a rapid assault. It is thought that Alzheimer's actually starts many years before symptoms show. The stage when first symptoms appear is referred to as 'mild cognitive impairment'. As Alzheimer's advances, the cortex withers and shrinks, causing problems with the brain's ability to plan, recall and concentrate. The disease also affects the hippocampus, which plays an important role in memory, and as this part of the brain shrinks the ability to create new memories is also hindered. By the final stage of Alzheimer's, damage is widespread and brain tissue has shrunk significantly.

On average, people with Alzheimer's disease live for eight to ten years after diagnosis. However, life expectancy varies considerably. How quickly a person's dementia will progress depends on many things. People who develop symptoms at a younger age often have a faster progression, and those with other long-term health problems such as a heart condition, diabetes or frequent infections may also suffer a more rapid deterioration. There are differences in speed of progression between different types of dementia; Alzheimer's has a relatively slow rate on average. However, it seems that most of the variation is between individuals. As with most conditions, genetics, environmental factors and overall physical health are all likely to play a role in the speed of disease progression.

Is brain volume lost in a similar way across different types of dementia – i.e. is this something intrinsic to dementia? The simple answer to this may well be no. A recent study revealed a fascinating finding that brain loss appears to be quite different in at least two types of dementia. This research followed 160 people who had been diagnosed with mild cognitive impairment. Of these, sixty-one developed Alzheimer's and twenty developed dementia with Lewy bodies (or Lewy body disease) during the course of the study. Lewy body disease is a type of dementia that shares a number of features with both Alzheimer's and Parkinson's disease. Whilst memory loss tends to be a more prominent symptom in early Alzheimer's than in early Lewy body disease, hallucinations and delusions, for instance, are more frequent in early-stage Lewy body disease than in Alzheimer's. In addition, motor and other physical symptoms, such as falls, a sudden drop in blood pressure, and urinary incontinence are more common in early Lewy body disease than in Alzheimer's. Brain scans over the course of the study showed that while over 60 per cent of those who developed Alzheimer's experienced significant brain shrinkage in the hippocampus, the vast majority (85 per cent) of those who developed Lewy body disease had a normal hippocampus volume. The people who had no shrinkage in the hippocampus were 5.8 times more likely to develop dementia with Lewy bodies than those who had hippocampal atrophy. The researchers suggest that observing shrinkage in the hippocampus may potentially be utilised as an indication of how mild cognitive impairment will progress in individuals; a lack of shrinkage over time may indicate that dementia with Lewy bodies, rather than Alzheimer's disease, is the likely prognosis. Treatments relevant to each condition may subsequently be better targeted at an earlier stage.

Although there may be some variability in the symptoms and pace of disease progression experienced by individuals, Alzheimer's typically follows a similar course in everyone affected. Early on, memory problems are a key sign that something is wrong. Indications may also include getting lost, repeating questions or taking longer to complete normal daily tasks. As the disease progresses, there is growing impairment of language and other cognitive functions. At first it might be increasingly difficult to find the right name or word, and later there may be problems with verbal and written comprehension and expression. Reasoning abilities, judgement and insight become affected, and there may be behavioural changes such as delusions, verbal aggression or wandering. Ultimately a person loses the ability to take care of simple tasks such as washing and eating and they may lose some motor control.

This regular pattern of symptoms is interesting. We know that in Alzheimer's, brain cells die in a fairly predictable pattern. What is interesting is that it has such a linear pattern at all. After all, this is not the case in some other conditions where both disease progression and the symptoms vary widely and are not easily predicted (such as MS). Why should it be that some diseases result in a brain response that is seemingly arbitrary whereas others follow a defined pattern? There may truly be a random element in some diseases, but perhaps also we just haven't identified all the patterns yet, or are not classifying diseases accurately. Whatever the case, it is clear that some diseases have a frighteningly clear path of destruction, forging through the brain rapidly and giving it little opportunity to defend itself. One such example is our next subject.

What have cows got to do with the human brain?

BEEF WARNING SPARKS PANIC, yelled one newspaper; MAD COW TRAGEDY BLAMED ON HAMBURGER, shouted another. Who could forget the terror instilled by the potential threat of mad-cow disease back in the 1980s and 90s? As we nervously ate our dinners, we looked on in horror at television images of burning cattle littering the British countryside. Were we all now at risk of a terrible fate? Was there really a ticking time bomb of mad-cow disease lying in wait?

Mad-cow disease, or bovine spongiform encephalopathy (BSE), is a neurodegenerative disease affecting the brains of cattle that can mutate and spread to humans. The human form is known as variant Creutzfeldt-Jakob disease (vCJD). Creutzfeldt-Jakob disease (CJD) is a rare, degenerative and fatal brain condition. It progresses rapidly and most patients die within a year of experiencing their first symptoms. It is one of a group of illnesses called prion diseases, which affect humans and animals. Prion diseases exist in different forms, all of which are progressive, currently untreatable and ultimately fatal. Their name arises because they are associated with an alteration in a naturally occurring protein: the prion. Prion proteins occur in both a harmless normal form, naturally found in the body's cells, and in an abnormal form, which causes disease. Once they appear, abnormal prion proteins clump together, and scientists believe that these clumps may lead to the neuron loss and other brain damage seen in CJD. However, it is not yet known exactly how this damage occurs.

CJD was first described in 1920, whereas the first UK case of vCJD was identified in 1986 and linked to BSE in cattle. In the UK there have been around 180 cases in total and it was concluded that vCJD was caused by feeding cattle the remains of infected cows, which then got into the human food chain. During this terrible phase in agricultural and public health history, it is estimated that 180,000 cows were infected – but in total 4.4 million were killed as a precaution.

There are actually three other types of CJD that are not caused by eating infected beef. One is a very rare genetic condition (familial CJD), and one is caused by the accidental spread of the disease from another person via contaminated surgical equipment or medical treatment (iatrogenic CJD). The most common form is sporadic CJD, where the specific cause is unknown. When we say common, it is still rare: worldwide it causes around 1–2 deaths per million population per year. In the US, for example, there are about 300 cases per year and there were 90 UK deaths in 2014 from this form of CJD.

There's no denying that CJD is an awful disease. In the early stages, people typically experience a loss of memory, and psychiatric symptoms such as depression or anxiety, behavioural changes, lack of coordination and visual disturbances. As the illness progresses, mental deterioration becomes pronounced and neurological signs, including unsteadiness, difficulty walking and involuntary movements, as well as blindness and slurred speech, may occur. By the time of death, patients have become completely immobile and mute. While some symptoms of CJD may be similar to those of other progressive neurological disorders such as Alzheimer's or Huntington's disease, it tends to cause more rapid deterioration of a person's abilities than most types of dementia.

Whilst cases of CJD are few and far between, and the outcome is inevitably the same, variability in individual response can still be observed. CJD usually appears in later life and runs a rapid course. Typically, the onset of symptoms occurs at about age sixty, and approximately 90 per cent of individuals die within one year. However, the variants of the disease differ somewhat in their symptoms and the course of the illness. For example, vCJD begins primarily with psychiatric symptoms, affects younger people (median age at death is twenty-eight) and has a longer-than-usual duration from onset of symptoms to death (a median of fourteen months as opposed to four and a half months).

Despite these sobering statistics there have been cases of individuals with CJD living much longer than the norm. Jonathan Simms was a talented youth footballer in Northern Ireland who first exhibited symptoms of vCJD, caused by eating infected beef, when just seventeen years old. Doctors gave him only months to live, but he went on to live for ten years with the disease. After a year and a half, and some lengthy High Court battles, he was given an experimental drug, previously only tested in animals, which appeared to stabilise and improve his condition for some time. Although we say he improved, Jonathan was still profoundly disabled. The role of the controversial treatment in Jonathan's disease is not clear and it is possible that he lived longer than others with the disease because of some inherent ability of his brain to withstand it longer. A clinician who treated him pointed out that he didn't start treatment until nineteen months after his symptoms began, which is already much longer than most people with vCJD survive. He may have had a natural survival ability beyond that of others.

Another person who lived longer than expected with vCJD was Holly Mills. She lived for nine years with the disease and was also treated with the drug that Jonathan was given. She too had already survived longer than average by the time she started treatment, perhaps also suggesting that she had a natural survival advantage.

Rachel Forber was a former soldier who was diagnosed with vCJD six months after showing signs of depression. She quickly deteriorated to becoming bedridden and requiring constant care, unable to recognise people or feed or dress herself, and only given a year to live. She was also given an experimental treatment (different from the one taken by Jonathan and Holly) and within three months she was able to get out of bed, walk unaided and swim without support. But the treatment didn't help for long: it eventually caused problems with her liver and she had to come off it. She went rapidly downhill and passed away a few weeks later. Was it the treatment that was effective, or that she was a very fit individual, or that she had a natural survival ability that enabled her brain to fight back to some extent before eventually succumbing to the disease?

Despite these examples of incredible fights against the disease, the fact remains that CJD is not a survivable condition. The brain cannot ultimately fend off the attack; the best it can do is to slow the disease down. Even if there were specific treatments, CJD would need to be conclusively identified in the first place. Unfortunately, currently there is no single diagnostic test for CJD and it can only be confirmed following pathological examination of the brain post-mortem. CJD causes unique changes in brain tissue that can be seen at autopsy: multiple microscopic and abnormal aggregates encircled by holes, resulting in a daisy-like appearance. For such a severe disease, causing immense damage to the brain, it is interesting to note that brain atrophy, although sometimes seen in cases of CJD, is not a prominent feature of vCJD. It is unclear why this is the case. It is possible that the typically very short duration between first symptoms and death with this disease is just not enough time for the brain to shrink in volume.

Eating infected beef can clearly have a profound effect on a person's brain, but it is by no means the only thing we consume that can have a lasting, and sometimes devastating, impact.

We are what we eat (and drink)

The brain can be attacked by things in our environment, and even things we do to ourselves. When exposure to natural or manmade toxic substances alters the normal activity of the nervous system, which can eventually disrupt or even kill neurons, we talk of 'neurotoxicity'. All sorts of things can cause neurotoxicity, such as radiation treatment, and exposure to pesticides, cleaning solvents or heavy metals, but let's take a look at one neurotoxic agent that has become the most socially accepted addictive drug worldwide: alcohol.

Research has shown that alcohol, or more specifically ethanol (the form of alcohol found in drinks), is neurotoxic, with direct effects on nerve cells. Once ingested, alcohol passes easily into the bloodstream and gets pumped all around the body. For many molecules, getting into the brain is a challenge as the blood-brain barrier protects the organ from potential harm from foreign substances. However, ethanol is able to pass through this barrier without any problems and goes on to alter the communication between brain cells. The brain is easily affected by the actions of alcohol, and heavy alcohol consumption has long been associated with brain damage.

Alcohol inhibits the function of neurons by reducing their ability to transmit electrical impulses. These electrical impulses carry information that is essential for normal brain function. By inhibiting the transmission of electrical impulses in neurons, alcohol can impair judgement, coordination, alertness, memory and visual perception, among other things. Although different parts of the brain are responsible for these functions, alcohol acts on the brain ubiquitously, and so any part may be affected. For example, the frontal cortex normally helps to suppress behaviours that are socially inappropriate and impulsive, and controls judgement and decision-making, but alcohol interrupts the balance and can result in impaired judgement, increased risk-taking and social disinhibition (i.e. potentially resulting in a person doing silly things while under the influence). The hippocampus is responsible for controlling learning and memory, and alcohol can prevent it from consolidating the information that forms a memory, leading to an inability to recall events during and after drinking (i.e. the person doesn't remember the silly things they did). These effects will sound all too familiar to anyone who has had a few too many drinks.

Beyond the occasional heavy night, there are of course many millions of people who regularly drink too much. Excessive alcohol consumption continues to be a significant problem throughout the world and is a leading cause of preventable death. The National Health Service estimates that around 9 per cent of adult men and 4 per cent of adult women in the UK show signs of alcohol dependence. This of course results in countless societal problems and an increasing incidence of associated health problems (e.g. deaths from liver disease have reached record levels, rising by 20 per cent in a decade), but what interests us here is the effect on the brain.

In the long term, excessive drinking leads to changes in brain function and can result in the organ shrinking. Evidence suggests the total brain shrinkage is probably due to a loss of both grey matter and white matter. The shrinkage occurs particularly in areas that are important for learning and memory, such as the cerebral cortex and the hippocampus. Many brain functions related to the frontal cortex appear to be affected, including personality and cognition.

However, the picture isn't all bleak. Long-term brain imaging studies have discovered that once people stop drinking for a sustained period, their brains increase again in volume. For example, longitudinal MRI studies of alcoholics have found that following just one month of abstinence from alcohol, cortical grey matter, overall brain tissue and hippocampal tissue increased in volume; and with longer-term abstinence there was a general increase in brain volume, particularly in the frontal and temporal areas. This suggests that the brain is able to recover to some extent from alcohol-related damage. It is thought that cortical white matter may be particularly amenable to recovery during prolonged abstinence from alcohol, although the mechanisms behind this are not yet fully clear. Reversal of brain damage doesn't necessarily happen for all recovering alcoholics, however, and there are a number of elements that are associated with a lower likelihood of recovery, including older age, heavier alcohol consumption prior to giving up, liver disease, malnutrition and smoking.

Alcohol-related brain damage is often mistaken for conditions like Alzheimer's disease. However, unlike Alzheimer's, it is not progressive and it doesn't inevitably get worse over time. Symptoms can improve significantly with treatment and much of the damage can be reversed.

One type of alcohol-related brain damage is a degenerative disorder called Wernicke's encephalopathy. People with Wernicke's may experience mental confusion, vision problems, hypothermia, low blood pressure, lack of muscle coordination and even coma. Wernicke's encephalopathy actually results from a lack of vitamin B1, better known as thiamin, which is vital for the growth, development and function of cells. Heavy drinkers often have low thiamin levels for a number of reasons: they may have a poor diet and may often vomit, thereby limiting their vitamin intake; alcohol can affect the stomach lining and reduce its ability to absorb vitamins from food; and alcohol can damage the liver, where thiamin is processed. Therefore, Wernicke's encephalopathy can actually be reversed by thiamin supplementation. If left untreated, however, this condition may go on to develop into a much more serious disorder called

Wernicke-Korsakoff syndrome, which causes irreversible brain damage. This syndrome is the best-known form of alcohol-related brain damage, although it is actually much less common than other forms, such as alcoholic dementia. The best-known symptom of Wernicke-Korsakoff syndrome is something called confabulation. This is when is a person can't remember the recent past, so they use the clues in the environment together with their long-term (intact) memories and knowledge to make up an explanation for where they are and what is going on. The person ends up with distorted, fabricated and false memories.

Other symptoms include amnesia, tremor, coma, disorientation, and vision problems. When left untreated, Wernicke's encephalopathy leads to death in up to 20 per cent of cases or to Wernicke-Korsakoff syndrome in 85 per cent of survivors. Wernicke-Korsakoff syndrome is found predominantly in alcoholics, although other causes of the disorder include nutritional deficiencies, eating disorders and chemotherapy.

Alcoholics who do not suffer from Wernicke-Korsakoff syndrome still perform worse than non-alcoholics do on tests of learning, memory, problem-solving, motor functioning and information-processing. (They are less accurate and take much longer to complete tasks.) However, much like the good news about the potential to recover brain volume, performance levels on many of these tests also appear to improve after several years of abstinence from drinking. Not all cognitive functions may return, though, and some individuals will have permanent impairments, particularly in memory and visual–spatial–motor skills.

Obviously, many people enjoy an alcoholic drink or two without seemingly experiencing any particular detrimental effects, but if consumption levels increase over extended periods, damage may start to set in. In fact, it is not just heavy drinkers whose brains may be affected. A recent study by Anya Topiwala and colleagues from the University of Oxford found alcohol intake to be associated with both a reduced volume of hippocampus and cognitive decline, even in moderate drinkers (who were three times more likely than non-drinkers to have hippocampal atrophy).

There is also individual variability in response to alcohol. Why are some people more vulnerable than others to the effects of alcohol on the brain?

Women seem more susceptible to the effects of alcohol than men in a number of ways. Women achieve higher concentrations of alcohol in the blood, due to having proportionally less water in their bodies than men, and become more impaired than men after drinking equivalent amounts of alcohol. Research has also found that women may be more vulnerable than men to alcohol-induced brain damage. Brain scans in one study revealed that the corpus callosum was significantly smaller among alcoholic women compared with both non-alcoholic women and alcoholic men, even when you take head size into account. When it comes to alcohol, there's no denying that men and women are different creatures.

A word on mental health

Whilst we have focused on physical health conditions in this chapter, of course a huge number of people will experience mental illness in their lifetime. Mental health problems constitute the largest single source of the economic burden from non-communicable diseases, with an estimated global cost of

$US2.5 trillion in 2010, predicted to increase to over $6 trillion by 2030.

The diversity of mental health conditions, symptoms and outcomes precludes us from delving deeply into this area. However, it is worth noting that whilst the world sometimes refers to 'mental health' and 'physical health' as separate things, there are in fact physical changes taking place within the brain during mental illness. The chance finding of brain shrinkage in people who had taken anti-psychotic medicine for schizophrenia prompted the writing of this book in the first place. We don't know what specifically caused the shrinkage, or what it means in practical terms for the individuals involved, but it is one of many studies that have found brain changes in people with mental illness. Much scientific research is taking place to try to understand more about what is happening inside the brain during different mental illnesses, what might be causing them and how to treat them most effectively. Many studies have shown that conditions such as clinical depression, anxiety disorders, schizophrenia and bipolar disorder cause measurable changes within the brain.

One fascinating project, led by the Stanford University School of Medicine, reviewed the findings from 193 brain-imaging studies, involving 7,381 people with mental illness. It found that there was a similar pattern in the loss of grey matter in the brains of people with schizophrenia, bipolar disorder, major depression, addiction, obsessive-compulsive disorder and a cluster of related anxiety disorders. Comparing the images with those from 8,511 healthy control subjects, the researchers identified three separate brain structures with a diminished volume of grey matter. Grey-matter loss in the three brain structures was similar across patients with different psychiatric conditions. These structures work together and are associated with higher-level functions, such as concentration, multi-tasking, planning, decision-making and inhibition of counterproductive impulses. The study also found that among healthy people, greater volume of grey matter was correlated with better performance in tests of higher-level functions. On the basis of these results, the researchers suggest that the grey-matter loss in the three brain structures is behaviourally significant, rather than just an incidental finding. Discoveries like this may pave the way for further research into understanding similarities, rather than merely differences, between mental illnesses, and even the potential for common targets for treatment.

Thinking about our loosely categorised 'cognitive conditions', we are not necessarily any clearer as to whether we can afford to lose any particular parts of the brain. With these conditions, the brain obviously prioritises basic functions that keep us alive and kicking over higher-level ones required for thinking and the like. This is clearly sensible in survival terms, so on that basis alone we can agree that parts of the brain involved in our core survival are more important than those that make us human. That's not to say that we don't need the other functions too, and we've seen several devastating examples of the consequences of not having all parts of the brain in working order.

Does the brain's response to disease reveal how much we really need?

The examples in this chapter provide a small insight into the fragility of the human brain, and also its ability to cope (to some extent) with sustained assault. Although we have also seen that the brain can fight back and recover, such as in remitting periods of MS, recovery from alcohol-related brain damage and even the extraordinary temporary partial recoveries of vCJD victims, the brain can only take so much, and with prolonged assault the neurons eventually start to die and the brain shrinks. Brain shrinkage seems to be a fairly common theme among chronic brain illnesses. But does this shrinkage mean anything in practice, or is it just incidental?

We know that shrinkage is not something particularly unusual that only happens when things go wrong in the brain. It is something that will eventually happen to us all, usually starting in our thirties. Our brains shrink as we age, particularly the frontal cortex it seems, but the detail of why has not yet been fully established. For example, we may lose neurons, our neurons may get smaller in volume, our white matter may decrease as myelin sheaths deteriorate, and/or there may be some organisational changes in the brain. Changes do not occur uniformly across the brain and the fact that different parts are affected to a greater or lesser extent is likely to account for the variety of cognitive changes we experience as we get older. The prefrontal cortex seems to be particularly affected, which corresponds well with cognitive changes seen in ageing, such as memory loss.

Beyond normal ageing, we have seen how many diseases result in abnormal brain shrinkage. There are many potential causes of brain atrophy, from neurological disease to brain trauma to alcohol and drug abuse. The truth is that for many brain conditions we don't yet fully understand the biological processes underlying them, which means there is a relative lack of effective treatments available and typically poor outcomes for those affected. The response of the human brain is hugely variable and we have an awful lot more research to do to understand why brain atrophy occurs, and whether and how it is affecting individuals experiencing such loss.

People are affected differently by the various neurological conditions, but if all humans have pretty much the same kind of brain, what is causing this variation? More women than men develop MS and dementia, but more men than women develop MND. Alcohol-related brain damage affects men more frequently than women, but women are more vulnerable to the effects of alcohol; when women are affected, they tend to develop alcohol-related brain damage at a younger age than men and after fewer years of alcohol misuse. Gender therefore seems to be linked with some of the variability, but there are different genes and environmental factors that also play a role, many of which we are still to discover.

We can't do much about our gender, and many things in our environment may be out of our control, but is there anything we can do to measurably improve our chances against a neurological attack? Well, we might be wise to do our best at school to get as good an education as possible and continue with lifelong learning. Research has shown that people with higher IQ, education, occupational attainment or participation in leisure activities have a lower risk of developing Alzheimer's disease. But the good news is that even if you didn't have a long formal education it may not be too late to take action. In one study of 128 people with MS, researchers found that whilst longer formal education was beneficial for reducing the degree of cognitive impairment over time, those with the most limited period of formal education but who had undertaken frequent reading, physical activity and challenging occupations actually fared the best. It goes to show that lifelong learning is not just beneficial for keeping up our interest in the world but also appears to have a physical effect on our brain, helping to keep it in good shape. What we're talking about here is typically referred to as our 'cognitive reserve': the ability of our brain to cope with potential damage and continue to function well.

Cognitive reserve seems to account for some of the variability seen between individuals in their response to neurological attack. The brains of those who respond better may have more efficient existing networks, have a greater capacity, be less susceptible to disruption, or be better at compensating for any disruption that occurs. In the case of natural age-related brain changes or Alzheimer's disease-related pathology, there is evidence that some people can tolerate more such brain changes than others and still maintain function. As we noted above, things like educational attainment and lifelong learning activities can increase this cognitive reserve to help the brain defend itself for longer. The more we understand about the role of cognitive reserve in protecting our brain, and how to improve it, the more targeted and effective we can be at designing interventions for future generations to carry out to keep the human brain healthier for longer.

Despite the potential benefits of increasing our cognitive reserve, we are still fairly defenceless against disease. In the previous chapter we saw how the brain has an amazing plasticity that can find ways to adapt to substantial acute change, such as from trauma or a missing piece from birth, yet it seems that, conversely, in the face of significant and sustained attack from disease it is still relatively poorly equipped to repair and recover.

Brain shrinkage, in different forms, is a key feature of many of the conditions we have considered, although we don't know whether it is a cause or symptom of disease. In some cases it seems that we can sustain considerable shrinkage before any symptoms become evident. Could this be an indication that in fact we don't need all of our brain? Does it mean we can afford to lose a bit of it before things get problematic? Is it a gross loss that we can sustain, or are there particular parts we can afford to reduce? We have a long way to go before science can tell us the answers to these questions. The bottom line is that we don't yet know if all this brain shrinkage is something to worry about. However, just because we can't yet say the shrinkage is specifically causing problems, it doesn't mean that we can assume that it'll be fine to lose a few per cent of our brain volume.

Let us now turn our attention to what we may be able to do to protect our brains from attack and consider whether there is any hope of a more resilient human brain, perhaps even a super-brain, on the horizon. We also need to consider how our brains evolve in the future? We're endlessly being told about the current obesity epidemic and how we all need to lose weight: maybe our brains are too big too? If future evolution of the human brain is to reduce in size or lose particular parts due to a lack of need, what part, or parts, do we think we could afford to slim down on? Part 4 of the book delves into the possibilities on the horizon.

A VIEW FROM

Ms Maggie Alexander, former Chief Executive of the European Multiple Sclerosis Platform, Brussels, Belgium, and former Chief Executive of the Brain and Spine Foundation, London, UK

Since training in neuroscience at the Wellcome Foundation, Maggie's career has been long and varied. She worked in biomedical publishing, information provision, campaigning and advocacy in a range of non-profit organisations focused on occupational and environmental health and safety, cancer and neurology. At her final role before retirement, as chief executive of the European Multiple Sclerosis Platform, she led a pan-European network of organisations in developing and implementing EU-wide programmes designed to maximise access to optimal treatment, care and research for the 2 million people in Europe affected by MS.

So, Maggie, in terms of what's happening in the brain, why do you think there is so much individual variability in people's experience of MS, even when they have the same type of MS?

'Whilst you may think of people varying in their experience of MS, it is not just about how things differ clinically but also how people differ in the choices that they make in response to their condition. As symptoms vary widely, therapy options also vary, and how people cope with their condition, and how they choose to approach its management, varies. Furthermore, across countries they have very different experiences of healthcare, what is available to them and how they are treated. In MS this may make a huge difference in the management and progression of an individual's condition.

'Of course, it is not just the course of the disease that is shaping a person's experience, but also how they adapt to it. For example, John Golding, inspirational past president of the European Multiple Sclerosis Platform, told me of how when he was first diagnosed with MS in his mid-twenties he became very depressed and suicidal. His MS progressed significantly and he reached the point where he was losing the use of his legs and falling and stumbling around like a toddler. He describes the day he got his wheelchair as one of the most liberating in his life because he could then move around independently. We often think of moving into a wheelchair as being a negative thing but for some people it is a real liberation and a positive adaptation to their disease. And the brain is likely to have also adapted in response; to both the change in mental well-being and to the change in physical requirements.'

Can you tell me about extraordinary examples from your experience of people adapting to significant levels of disease?
'Of course, I've met many, many absolutely extraordinary people! One springs to mind: do you remember the boxer Michael Watson? He was very nearly killed in his world-title fight with Chris Eubank. [He famously suffered severe and near-fatal head injuries in the 1990s and that he survived was considered extraordinary.] Neurosurgeon and founder of the UK's Brain and Spine Foundation Peter Hamlyn practically rebuilt him. I was told by Peter that the chances of Michael surviving were very, very slim. However, in 2003 he managed to walk the London marathon in six days [an effort described by Peter Hamlyn as amounting to '12 years, 6 operations, 3 hospitals,

26 miles and 385 yards']. Whilst he never made a full recovery and still requires the help of a carer, what he has achieved in the face of extreme brain adversity is phenomenal. He was very inspiring. Presumably a combination of his personality, religion, family and medical team, amongst other things, helped achieve his remarkable recovery. An enriched environment is so important for the brain. It is about people and relationships.

'Sue Tilley is someone I know who I think is another fascinating example. She has had relapsing-remitting MS for many, many years but she has been virtually symptom-free for thirty-three years. Of course if she gets very tired or ill she takes longer to recover than other people, and she has a slight weakness in one leg, but otherwise she stays well. Now, the majority of people diagnosed with relapsing-remitting MS go on to develop secondary MS [the next stage of the disease after relapsing-remitting MS], so why not Sue? Is it a combination of the person, their genes and their environment, or is it some sort of unknown biological component that has yet to be identified? It is fascinating that some people do not experience disease progression.'

Bearing in mind the many people you have met, how much brain do you think humans really need?

'I don't think it is so much about the amount of brain but more about ensuring as a society that the bits of the brain that are really important are nurtured. I do think that the better your skills are in relation to people and social situations, the easier the passage through life, so perhaps those parts of the brain related to these functions are some of the most vital.'

Maybe, as Maggie suggests, other than those parts that regulate essential physiological functions (such as breathing), the brain really needs to preferentially protect the parts involved with social interaction and communication. Perhaps ultimately that is what makes us human.

PART FOUR

Future Perfect

Can we enhance our brains?

CHAPTER 8

The Optimised Brain: How much better could the human brain be in our lifetime?

The 2016 summer Olympics saw an astounding twenty-seven new world records and ninety-one new Olympic records, as athletes ran and swam faster, threw further, lifted heavier weights and propelled their bikes, boats and canoes more quickly than ever before. The swimmer Michael Phelps won his twenty-eighth Olympic medal, making him the most decorated Olympian of all time. Usain Bolt won three gold medals for his third successive games: the fastest human ever to run 100 or 200 metres.

Perhaps we should not be surprised that the world's top athletes, through fair means or foul, get better on average each year. The 2016 generation of Olympians benefited from the most sophisticated training, nutrition and recovery regimes that have ever been devised. Being the fastest human alive today brings more glory, fame and financial reward than ever, so anyone with the raw talent is probably more motivated to succeed (and more likely to access training) than ever before. And since more and more humans are born on the planet each day, the chances of a new record-beating athlete being born must statistically increase each year. With this combination of statistical probability and scientific advance, we should perhaps expect world records to continue to fall for ever – or at least for the remainder of the century.

In Chapter 4 we noted that brain function, at least as measured by IQ tests, seems to be improving with each generation, just as markers of physical health, such as average height and lifespan, are. There's no universally recognised brain equivalent of the 100-metre sprint, so we simply don't know who holds the current world record for mental agility. But we can assume that as the world's population as a whole gets smarter, the people at the very top end of the distribution do too. And that means that, just as with sprinters, there's a good chance that the smartest person who has ever lived is alive now.

Let's consider what the best human brain currently available might be like. Here and now, just how good could a human brain be, and is there anything we can do to achieve it?

Getting a head start

The most important thing you can do to ensure you have an optimally functioning brain is to pick your parents wisely.

We have already discussed how children's brain function, like their height, weight, skin colour and food preferences, is determined by a combination of genetic and environmental effects: by nature, nurture, and interactions between the two. We can roughly work out the relative contributions of genes and environment by studying the extent to which traits run in families, and particularly by looking at the degree of similarity of traits in identical twins (who share 100 per cent of their genes) versus non-identical twins (who have 50 per cent of genes in common). These kinds of studies find that, on the whole, brain 'hardware' traits tend to be highly heritable, with perhaps 75–90 per cent of traits such as total brain volume due to genes. In contrast, brain 'software' traits, like intelligence and personality, are a bit less heritable, with about half of the variation in a population attributable to genetic factors.

In contrast to something like eye colour, which is determined by a relatively small number of genes, intelligence is influenced by common variations in thousands of genes. Thankfully, this means that the idea that we might genetically engineer super-smart babies is likely to remain in the realm of science fiction for a while yet, because trying to artificially engineer an optimal set of gene variations would be a seriously complex procedure. Luckily, however, for those keen to increase the odds of having highly intelligent babies there's a simple way of doing this already, and that is by selecting a highly intelligent mate.

At least half your brain function throughout life will be determined at conception by the DNA you inherit. Paradoxically, it turns out that the relative influence of genetics versus environmental influence on intelligence changes across a lifetime, with genetics becoming increasingly important through childhood, adolescence and adulthood, even though the amount of environment experienced also increases. Take identical twins, for example: one might imagine that their IQs would be highly correlated in childhood, because factors like diet, education and life experiences in general would be more similar then, and less correlated in adulthood once their paths diverge. Strangely, the opposite pattern is seen: the IQ scores of identical twins get more similar, not less so, as the twins get older. Indeed, the effect of genetic influences on IQ in all of us increases throughout life, probably until at least the age of seventy.

This isn't intuitive: you might instinctively think that babies come out of the womb a blank slate, entirely the product of their genes at this stage when so little experience has occurred. Why is this not the case? One reason is that environment doesn't just 'happen' to us: our genes themselves play a role in selecting the kind of environment we spend every moment of our life in. And over the course of a lifetime this builds up a greater and greater 'dose' of environment; hopefully one that suits us best.

To illustrate this, let's think about a boy who has an inherent musical talent. He is likely to have had musical parents, and to have inherited genes that predisposed him towards musicality. But these parental tendencies will also have resulted in him being exposed to more music from an early age than other children. As a baby he would have had little choice about this, but every parent knows that even very young children have ways of making their likes and dislikes known pretty early on. So if as an infant and toddler he seemed to enjoy music, he might be taken to music-based pre-school activities, and then be encouraged to take lessons and attend concerts more than a non-musical child. Since musical children usually turn into musical adults, across a lifespan there may be enormous differences in the number of hours of exposure to music that an inherently musical baby will eventually experience. As we see in this example, one of the reasons it can be so hard to tease apart genetic and environmental effects is because the two often work together.

Another reason it is difficult is because individual genes can simultaneously affect multiple different aspects of health, a phenomenon known as pleiotropy. One of the largest studies to look at this in humans is the UK Biobank, a huge ongoing study of middle-aged adults who have undergone all sorts of medical tests, and agreed to have their health followed in the future. Studies of the genome of more than 100,000 Biobank volunteers have shown that the sets of genes that have been found to drive variation in their cognitive functions overlap significantly with the sets of genes that drive other diverse aspects of mental and physical health. For example, there is significant overlap between the list of genes that influence verbal and numerical reasoning skills, a general measure of adult IQ, and those that influence (among other things) intracranial volume, body mass index and risk for ischaemic stroke (the result of an artery that is bringing blood to the brain getting blocked). And the list of genes that are associated with educational attainment, a measure that is partly but not entirely correlated with IQ, were in addition found to overlap highly with the genes that predict risk for coronary artery disease.

How can variation in one gene affect both brain development and risk for cardiovascular disease? To answer this, we again need to move beyond the first steps of inheritance, and start thinking more about other things that influence the brain as it becomes an ever more complex and experienced organ.

The importance of the womb

As well as choosing your genes wisely, you'd need to choose carefully the womb in which you spend the first few months. This environment is the first and most important that the developing brain experiences, and it's no exaggeration to say that the environment a foetus experiences during pregnancy has effects that last a lifetime. But before any mums-to-be rush off to play Mozart to their bump, let's explore a little further what we think might matter when in the womb.

In the 1980s, a researcher named David Barker was studying variation in the nutrition of mothers in different parts of England. He noticed that babies who had a lower than average birthweight went on to have a higher than average risk for heart disease in adult life. He went on to propose the theory that, during pregnancy, the foetus learns a little about the world it can expect to be born into, and adapts accordingly. For example, if a mother has poor nutrition during pregnancy, the foetus undergoes physiological and metabolic changes that will prepare it for a world where nutrition is going to be scarce. If that baby then grows up in a world where sugar is readily available, its prenatal programming will make it particularly at risk for developing type 2 diabetes in later life.

When Barker first published his theory it was met with a fair degree of scepticism. It seemed so unlikely that those first nine months of life could have such a large influence on health outcomes many decades later. One fellow epidemiologist, Janet Rich-Edwards from Harvard, was determined to prove it wrong, and could access the birthweights of more than 100,000 nurses whose health she had been following for many years. To her astonishment she found a strong replication of Barker's hypothesis: the lower the birthweight of the nurse, the more likely she would later have a heart attack or stroke. Many other studies have since confirmed the effects on many adult metabolic diseases, and related markers such as blood pressure and insulin resistance.

So childhood and adult metabolism is definitely affected by the environment that a foetus experiences. What about the brain? It turns out that low birthweight predicts lower IQ throughout life, including slower cognitive development in childhood and more rapid cognitive decline in late life, and other brain health outcomes, including increased risk for later depression. These effects are big enough that they should interest all of us, not just statisticians: a baby who is born at 5.5 pounds (2.5 kilograms) or less will likely score around 5–7 IQ points lower in adolescence and young adulthood than they would otherwise have been expected to do, and have twice the risk for adult depression.

We can think of these early influences on the brain as operating at two levels. One is that birthweight is simply a biological marker for how optimal the womb environment has been for that foetus. A low birthweight suggests that things haven't been perfect so far – and brain development too might not have been quite as good as it otherwise could have been. We know that other general markers of physical development also predict how robust a brain is – for example, people who have longer limbs and are taller have a slightly lower risk of developing dementia. This is thought to be because longer limbs or increased height are a sign that early physical development, including brain development, was just a little better. This may be one effect of the pleiotropy we mentioned before: for example, a gene that affected the foetus's ability to harvest adequate nutrition through the placenta might simultaneously affect the early growth of many different organs.

We should point out that these associations reflect only average differences and the effects are small: clearly you can be born smaller than average, or be a short or short-limbed adult, and yet do brilliantly well in life. The effects are important and interesting because of what they tell us about how early biology affects the brain, not because they tell us anything useful about how well any one individual will turn out.

But it is important to think about how foetal environments come to differ in the first place. This almost certainly involves a big mix of genes, environment and plain random chance – very little of which a mother could control. Even so, expectant mothers today are bombarded with advice about how to keep their foetus healthy: to avoid smoking, alcohol and certain high-risk foods, while making sure to take prenatal vitamins, eat healthily, exercise and avoid excessive weight gain. Some of these messages, such as the dangers of smoking during pregnancy, are almost universally known – plastered as they are across every cigarette packet in the UK. Yet whilst some mothers follow such advice rigorously, others don't.

It's reasonable to think that there may be many differences between mothers who do and don't follow advice to stop smoking when they become pregnant. These may be differences in income, education level, family history, mental well-being, or tendency to addiction. Indeed, all of these things also predict the likelihood that someone will be a smoker at the time that they get pregnant. And since smoking in pregnancy definitely slows foetal growth, that means any or all of these risk factors are likely to be associated with differences in babies' birthweights.

A pregnant mother's healthy and unhealthy choices don't happen at random; they are strongly tied in with her environment: her upbringing, socioeconomic situation, education and ability to look after herself. And since these are factors that don't change quickly, the baby will most likely be raised in the same environment as its foetal self was. So when we note that babies with low birthweight tend to grow into children with lower IQs, we could just be saying that if you experience tough circumstances when you are a foetus, you are more likely to experience them as a baby too, and throughout childhood.

We cannot ethically randomise babies to receive suboptimal environments. But if you study very large numbers of people, and use clever statistical analyses, it is possible to pick apart the effects of ongoing circumstances from the kind of biological pre-programming that Barker was talking about. (Indeed, if you're going to make claims that something causes something else, you need to be very good indeed at accounting statistically for all the other possible explanations.) From the best studies that have been conducted so far, it seems that early brain development is indeed affected by both prenatal environment and myriad factors later on in life. So, assuming you picked your genes well, and had an optimal experience before being born: now what?

Enriched early years

The third reason to pick your parents well is because they have a great deal of influence over how you spend the early years of your life. Indeed, modern parenting can sometimes seem like a minefield of loaded decisions, each of which might, in some often unspecified way, hugely influence your child's cognitive development and future well-being. Does it matter if you have a few drinks before breastfeeding? What about sleep, and potty training? Organic baby food? Is it really worth getting everyone dressed and out the door to attend baby yoga and swimming lessons, toddler gymnastics or after-school music or language classes? If you did none of those things, would your child be any worse off in life? It turns out that, for the majority of these decisions, for the majority of people, the realistic answer is: well, it's not a bad idea, but only if it's not too inconvenient.

To give a more scientific answer, we need to think about two different classes of environmental influence, one of which has direct biological effects, and another which we might term 'enriched environments'.

Let's take a biological example first: breastfeeding. For most of human history breastmilk has been more nutritious and safer than anything else you could feed a baby. Only in the last few decades have nutritionally balanced formulas been available that supply as good a source of energy, protein, vitamins and minerals as human breastmilk. Even today, in parts of the developing world, even if such formula is available, safe water with which to mix it and the ability to sterilise bottles and other baby-feeding paraphernalia are not. In addition, colostrum, the thick yellowish milk produced immediately after labour, contains antibodies that protect against dangerous illnesses such as diarrhoea and flu. With these two very good reasons behind it, the World Health Organization recommends that babies are exclusively breastfed until six months, and continue to receive breastmilk until the age of two. Current estimates are that around 800,000 lives could be saved a year through optimal breastfeeding.

The environmental factors that influence breastfeeding choices in the developed and developing worlds are clearly quite different. In the US, for instance, many women return to work long before the six months of recommended exclusive breastfeeding is up, and for this and other reasons many mums find it preferable or more convenient to stop breastfeeding, if they ever started, earlier than this. From those who wish to breastfeed but are unable to, to those who wish not to but feel pressured into it, it can be a hugely emotive topic for many parents. But from a scientific point of view, we do know that even in developed countries where safe alternatives to breastmilk are available, breastfeeding does seem to confer an IQ advantage on babies.

Our best guess at the moment is that breastfeeding a baby boosts its later IQ by about three points. Just like with our earlier discussion of maternal smoking, it can be extremely hard to pick apart the effect of breastfeeding from other factors likely to go with it – including the mother's IQ (which will directly influence the child's IQ) and socioeconomic and educational factors (which will be correlated with both the mother's IQ and the likelihood of her breastfeeding). The best evidence we have comes from studies that try to statistically control for all of these influences. While it's not ethical to randomise mothers to breastfeed or not, it is ethical to randomise mothers to receive extra breastfeeding help – and one study that did that in Belarus showed both an increased rate of breastfeeding and an increased IQ in the babies of mothers who were given extra support. Altogether this gives us reason to think that breastfeeding provides a small but real boost to childhood IQ. Whether this lasts throughout life is even more difficult to work out, but since childhood IQ predicts adult IQ, and later educational and economic outcomes, we can assume that at least some effect of this early boost will transfer through to adult life.

What about social or cultural aspects of the baby's environment? Does hearing Mozart rather than, say, pop music prime the baby's brain to expect a higher level of stimulation, perhaps giving a head-start on early cognitive development? The short answer is that within the range of what one might consider reasonable ways to raise a child, who you are as a parent matters much more than any of the child-rearing decisions you actively make.

Let's break this down. First of all: who you are. Your genes matter. As we discussed earlier, somewhere around half of normal variation in traits we are interested in, like brain size, IQ scores and mental health, can be attributed to genetic factors. And we know that the parents' educational, social and economic situation matters a great deal too – so in any discussion about children's outcomes we have to take these into account. What doesn't matter much is the choices you make for your kids. We're not talking about abusive or neglectful circumstances here, which can clearly have very long-term and serious effects. But there is little evidence of any long-term cognitive benefit, or detriment, coming from the day-to-day choices of well-meaning middle-class parents. Whether to take them to music lessons, or swimming, French classes or Bible study. Whether to feed them organic, gluten-free, home-cooked meals or microwaved TV dinners. Whether to let them watch cartoons or play violent video games or encourage them to read classic works of literature. Of course these choices do have specific effects – on their ability to impress a future girlfriend by ordering a meal in French, say, or debating the finer points of the plot of *Pride and Prejudice*. But there is little evidence that they have any effect on how well, overall, a child will do: how happy they will be, how much money they will earn, how long they will live. All of these things are driven overwhelmingly not by parenting choices but by two other things: who their parents are, and the specific things that happen to them in life.

How do we know this? Well, in large part the same way we know how heritable any of these things are: by looking at similarities and differences between family members with different amounts of shared genes, and different amounts of shared environments. To a geneticist, everything 'environmental' (that is to say, not genetic) is either shared or unique. Shared environment is that which is shared, for example, by brothers and sisters within a family. If your parents have liberal views about homework or recreational marijuana use or make you go to bed early or feed you exclusively on fast food, that's all part of your shared environment. Unique environmental influences are the things that happen only to you and not the rest of your siblings: the bang on your head you got when you were five, the sweets you used to sneak off and buy with your lunch money, the teacher you had who got you really excited about chemistry when you were thirteen. To put it another way, shared environment is the stuff that parents can affect; unique environment is the stuff that they can't. And the large number of studies that have been done in various interesting family combinations – in twins reared together and twins reared apart and adoptive families of biological and non-biological siblings – all give the same answer: shared environment has a small effect compared with both genetics and unique environment.

What is it in the non-shared environment that affects how kids develop? One major factor, particularly when it comes to how children behave, is peer influences. Surprisingly, who a child goes to school with, and chooses to hang out with, has a far bigger socialising effect than their parents, since children learn how to interact socially from their peers. On the whole, children, and especially teenagers, select peers who are like them in some way, and then do what their peers do, dressing how their peers dress, etc. Another thing that's worth noting is that chance, or randomness, also falls into the category of non-shared environment and that, of course, can sometimes play a big role.

Becoming who you are

The other thing that drives what children become is what they are early on in life. We discussed this a little with respect to musicality: that children's parentage can simultaneously be influencing them and influencing the environment that they are exposed to, and that this is cumulative across a lifetime. So some of the things we think of as being purely environmental may operate partly or mostly through genes. And some of the genetic drivers of healthy and unhealthy outcomes may operate through behavioural choices.

Let's take a case where there is a really clear association between a behaviour and a disease: smoking and lung cancer. There is absolutely no doubt that smoking, an environmental factor, increases risk for lung cancer. But there are genes at work here too: in vulnerability for lung cancer, but also in the tendency to smoke, and to continue smoking even when you know the negative consequences. Some people find smoking more pleasurable than others, some find it harder to give up than others. And some of that variation is due to genetic factors.

We believe that many of the pathways between genes, early environment and brain health throughout life operate through similar behavioural choices and tendencies. But it can be hard to know which comes first; the risky behaviour or the health state. For example, we know that people with schizophrenia use cannabis much more than the general population. But which causes which? Do people with the early symptoms of schizophrenia use cannabis to help them deal with those symptoms, or does cannabis use predispose to, or trigger, psychosis? (For what it's worth, researchers think cannabis use does make the brain more vulnerable to developing schizophrenia, but it is hard to be sure.)

The best tool we have in our arsenal to answer questions like this are research studies that follow people across a whole lifetime, preferably starting before they are born. One such study enrolled more than 1,000 babies born in the town of Dunedin, in New Zealand, in 1972 and 1973. Now in their forties, 96 per cent of those babies still take part in the study, and this has allowed the Dunedin cohort to teach us a lot about the effects of early environments on the brain, and how these play out in both those whom life treats well and those who hit many problems along the way.

One startling thing the Dunedin study has taught us is about the lifelong effects of differences in early temperament, specifically in self-control. Children who are high in self-control tend to be conscientious: they have learned to control their impulses and emotional outbursts to some extent, and they are able to delay gratification. These types of behaviour are regulated by the prefrontal cortex, which, as you may remember from our discussion of ageing, is the part of the brain that develops last, typically continuing to mature right through adolescence and into the early twenties. So children with self-control have, on the whole, a more functionally mature brain and, like other forms of temperament, the tendency to high or low self-control is one that tends to stick around throughout life. It turns out that this early tendency has lifelong consequences: the Dunedin children who had poor self-control (as reported by researchers, teachers and parents) from ages three to eleven, had a much higher risk for a range of negative outcomes in later life. These include poorer financial status (lower income, savings and home ownership), poorer physical health and a higher likelihood of being a single parent, having a criminal record and using drugs.

Records of how the Dunedin participants behaved as teenagers help us to explain some ways in which children who had poorer self-control in their early school years ended up as worse-off adults. During their teenage years, they were much more likely to get caught up in troubles like dropping out of school early, smoking, and becoming pregnant. And genetic analysis has shed further light on the drivers of success in the Dunedin babies. Those who were born with a more 'successful' set of genes (genes that had, in other studies, been shown to be associated with a higher educational status) tended to do better throughout life, from learning to speak and read earlier, to gaining higher-status jobs and partners in adult life, to better retirement planning. Children with a higher genetic propensity to success were also more socially mobile, regardless of the social class in which they had been born. By studying so closely the events of these children's lives we can get a glimpse of the myriad ways in which differences in certain brain functions can affect just about every aspect of our later life.

Brain help for adults: what you can do

By the time you are reading this, your early development and even your teenage choices are likely to be a done deal (if not, then well done you for being so ahead of your peers!). So what can we do now to optimise our brains, and protect them against the ravages of ageing?

There are plenty of suggestions out there, but considerably fewer that have any weight of scientific evidence behind them. Here we consider the two that have been best researched, and about which we can say with reasonable confidence: this will help. As with physical health, there is no magic bullet to preserve and protect the brain. The two things we can do are to be physically active and stay mentally active. When we talk about healthy brain ageing we are really discussing one of two things: how to minimise ongoing damage to the hardware of the brain, mostly by keeping its blood supply as good as possible; or how to optimise the operation of the brain's software.

There is plenty of evidence that people who stay mentally active, such as learning a new language or a musical instrument, doing crosswords or taking part in other intellectually challenging activities, preserve full cognitive function for longer in life. We need to be very careful here, however, about the direction of causality. It may be that people who are cognitively intact get more pleasure from cognitively challenging activities than people whose faculties are starting to fail. For this reason, it is difficult to run methodologically rigorous studies to test the effectiveness of things like brain-training programmes, which use a regime of increasingly challenging but enjoyable puzzles or games that are designed to build up people's cognitive function (rather like lifting increasing weights to build up muscle strength). Because people will choose, and then adhere better to, a regime of activities that they find more enjoyable, it is hard to run a well enough controlled study to decide whether any particular brain-training package or cognitive activity is really actively supporting healthy brain function.

What is clear is that people who have spent more time doing cognitively demanding activities over a lifetime are to some extent buffered from the physical effects of brain ageing and degenerative diseases such as Alzheimer's. As we noted in Chapter 7, we call this buffer 'cognitive reserve', the idea being that it is a back-up reservoir of brain function that protects us from the functional consequences of normal or disease-related brain damage. That cognitive reserve exists is now quite clear: people with higher IQ, more years spent in education and more cognitively challenging employment histories have a lower risk for dementia, even though their brains show normal amounts of age- and disease-related damage. In fact, post-mortem studies have shown that people with higher cognitive reserve who do get dementia show less severe symptoms, even when they have a greater amount of brain damage than those with lower cognitive reserve.

We think cognitive reserve can be built up right throughout life, so taking part in cognitively challenging activities, learning new skills and continuing to 'use it or lose it' probably applies no matter how old you are. However, for many people, life is already cognitively challenging, especially for those in higher education or knowledge-based jobs. So in between making great teenage choices and staying intellectually engaged in retirement, what is the best thing you can do for your brain?

There's no way around it: the best thing you can do to support your adult brain is to stay physically active. There's a fairly simple explanation for this, which is that the brain uses a great deal of the oxygen and energy that the heart pumps around the body. Poor cardiovascular function, and associated symptoms such as the build-up of deposits in the arteries, leads to damage to the brain by preventing an adequate flow of oxygen and other nutrients. This can cause considerable chronic damage over time (analogous to cardiovascular diseases such as coronary artery disease and heart failure), and it can also cause acute problems such as stroke (equivalent to a heart attack). Retaining good cardiovascular health helps maintain good condition in all the bits of brain machinery that support cognitive function.

'Aha,' you might say, 'but surely most of this evidence is observational, and so might suffer from the same confounding elements as we have tackled repeatedly in this chapter. People with a higher education, or IQ, or better social or financial situation, may be more inclined to take care of their health in other ways too: they may have a better diet, be less stressed, be less likely to smoke, and more likely to go the doctor at the earliest sign of a problem. Mightn't all these things affect brain health too?'

They might indeed. And there are certainly plenty of limitations in studies that depend on asking people how much they exercised last week – let alone across a lifetime. As so often, we'd consider a randomised controlled trial (RCT) the strongest test of whether exercise really improves your cognition, or your chance of developing a cognitive disorder. In an RCT, participants are randomly assigned to an intervention or a control condition. In an RCT of a medication, the intervention is the medication and the control condition is often a placebo. When the intervention is physical activity, this is not always as simple as it could be. Aside from deciding what, and how often, and how vigorously your participants should be exercising, how can you ensure someone who normally takes little exercise and then is randomised to take more will actually do so? Should the control condition be 'Exercise as you usually would' or 'Don't exercise'? How long would an intervention need to go on for to be expected to make a difference compared with the impact of a whole lifetime of prior behaviour?

One good place to look for evidence is situations where the conditions for physical activity can be imposed by some higher power. Whilst military training, or perhaps a prison system might work, they are not likely to be typical of most lifestyles. Instead we turn to children.

There have been quite a few attempts to run exercise RCTs in children and young people, most frequently by changing the amount of mandatory sports or physical education (PE) lessons that children take part in for the duration of the study. The studies tend to be small, and over relatively short timescales, but the good news is they do find small improvements in academic achievement and/or cognitive performance. Because increasing the amount of scheduled PE in the school week is likely to be quite an effective way of ensuring an increased amount of children's physical activity, these studies certainly support the story we are building of exercise being a major driver of optimal brain health.

We already know that exercise is good for us, and yet most of us still don't do enough, perhaps because as a species we are quite bad at enduring short-term pain for long-term gains. If we're interested in shorter-term effects on brain function – whether it's worth going for a run when you're studying for an exam, say – we need to be asking about what happens in the brain within minutes of exercising (which we detail shortly), and whether that helps or hinders cognition.

One (psychological) explanation is that exercise increases mental arousal and alertness, and thereby improves our ability to process information. A more biological explanation is that exercise releases a cascade of chemicals, such as dopamine and adrenalin, some of which are key drivers of the cellular processes that have to occur in order to form a new memory. In other words, acute exercise primes the brain at a molecular level to better process memories and other information and, regardless of how much exercise you have taken over a lifetime, exercising today will help your brain be best prepared to lay down new memories today.

So there are positive effects of exercise on brain chemicals. However, if you suddenly start slogging away on a treadmill for several hours a week, what's going on in your brain is likely to be the least obvious of the physiological changes that occur. Yet what's going on in your muscles may also affect your brain function. A recent set of studies in mice, monkeys and rather sedentary university students has pointed the finger to a chemical called cathepsin B, a protein secreted by muscles when you exercise. When previously sedentary students went through a tough four-month programme of treadmill running, the increases in their blood levels of cathepsin B correlated with the extent to which they improved on measures of visual recall, such as the ability to draw from memory.

How could something secreted by muscles help these students perform on a cognitive test? The answer lies in the hippocampus. One reason we keep returning to this brain structure is because it is one of only two major areas where new neurons can be generated. This is pretty remarkable: up until recently it was thought that all the neurons you would ever have were formed during embryonic development, but it is now estimated that the adult human hippocampus generates around 700 new neurons a day.

Unfortunately, there's no good way to accurately measure day-by-day changes in the living human hippocampus; but we do know that when rodents are housed with running wheels, exercise increases both their cognitive abilities and the number of new neurons that develop in the hippocampus. This is likely to be the major mechanism explaining how exercise increases the size of the hippocampus and boosts memory. The process of neurogenesis is regulated by chemicals called growth factors, and cathepsin B appears to regulate some of these growth factors. It's quite a journey that it goes on: generated by muscles, travelling through the blood and across the blood-brain barrier and eventually finding its way deep inside the brain, where its presence stimulates new neurons into being.

One last thing about physical activity: because the immediate and the long-term effects of exercise on the brain are quite different, they represent two independent strategies for optimising brain function. What if you put them together? One recent study tested this by randomising seventy-five young adults who were not normally physically active to different levels of exercise for four weeks. Their memory was tested at the start and at the end of the four-week period. As expected, those who had exercised more during the four-week period showed the greatest gains in memory – but the group who performed best were those who had also exercised on the day of testing.

Perhaps the next time you sit on your sofa thinking about how you really should go for a run; knowing that you're investing in today's brain function as well as tomorrow's will help get you out of the door.

What limits our brain function?

None of us can choose our parents, nor the experiences of our early years. Even our parents' best intentions have little effect, it seems, on the development of the optimal brain. However, in this chapter we have discussed two adult behaviours that can do such a thing, and we might have discussed more: how stress is generally bad for the brain, how managing that (by meditation, yoga, music, socialising or many other means) is good for the brain. We might have dwelled more too on the effects of sleep that we discussed in Chapter 5 and concluded that this offers a daily top-up of brain power that we would do well to encourage. But none of this advice would come as a surprise. We already know what is good for our physical health, and if we have learned one thing here it is probably that brain health is really no different. Is it worth changing your life to try to optimise brain function? Sure it is. And you should do so knowing that we are creatures of habit, and that the lifestyles we adopt, intentionally or otherwise, tend to become lifelong patterns of behaviour.

Yet, despite all our best intentions and those of our parents, we do still vary in brain function and all that that brings: in educational attainment, career and social success, the preservation of cognitive abilities in later life, and in happiness. It is perhaps these last two goals that we should pay most attention to. It is ironic that we know far more about what leads to mental illness and cognitive decline than what leads to mental health and successful cognitive ageing. Only recently has research interest begun to shift towards people who seem to have optimised brains: those who remain cognitively sharp into extreme old age, and those who have excellent mental health despite enormous challenges. These people are rare, but they exist. What can we learn from them about the limits of human brain function?

One particularly interesting attempt in this field is the Northwestern University 'SuperAging' project in Chicago. This is an ongoing research study into people who are over eighty but have cognitive capabilities (particularly memory) no different from those seen in healthy fifty- and sixty-somethings. Neuroimaging studies of the SuperAgers' brains show that they have a thicker cortex than other people of their age, and the same amount of intact brain tissue as people in their fifties and sixties. In the right anterior cingulate, an area associated with cognitive and emotional processing, the SuperAgers have a thicker cortex than even volunteers several decades their junior. In fact, in the brains of these youthful eighty-somethings key brain networks that control functions such as memory and attention look just like those of young adults.

So people who are exceptionally cognitively intact in their old age are probably benefiting from brains that are in some ways age-resistant. But what do we know about people with exceptionally good mental health? In Dunedin, only 17 per cent of the study group would not, at some point in their first four decades of life, meet criteria for at least one mental illness. This makes an important point: that lifelong mental good health is rare and that periods of poor mental health are the norm for most of us. What characterised that minority who did show persistent good mental health was not any of the things you might at first predict. They were not born into wealthy families, did not have particularly good physical health or exceptional intelligence. Instead they seemed to have an advantageous temperament, and no family history of mental illness – from which we might infer that they are lucky in both their genetic and early environmental backgrounds.

So how good could a brain be right now? Given particular fortune in both early development and resistance to age-related decline, a brain might reach peak function in its late teens or early twenties and could, if lucky, stay that way for sixty years. But the brains that are on the planet today have more opportunities than ever before: optimal nutrition and education; limitless access to the existing store of human knowledge through the internet and other information technologies; and less poverty, drudgery and intellectually undemanding work than ever before. Today's SuperAgers, for all their luck, have nothing compared with the technological and scientific advantages that their grand- and great-grandchildren will be born into. In Chapter 9 we step away from our careful consideration of the state of science today, and hazard a guess as to how a future brain might look different from yours or mine.

CHAPTER 9

Coming Soon to a Brain Near You: Can we preserve brain power and even enhance it via artificial means?

As a race, humans are willing to go to extraordinary lengths to enhance their beauty. Globally, we spend $180 billion a year on cosmetics. In 2014, the American Society of Plastic Surgeons provided more than a quarter of a million women with breast implants, earning more than a billion dollars in the process. How far, then, might we be willing to go to enhance our brain function?

What if you could raise your IQ, increase your motivation or intensify your charm with just the click of a button, the swallow of a pill or the zap of an electrical current? Ideas like this have, over the years, been the basis for some great science-fiction plotlines. In this chapter we consider how close they are to being reality.

We also consider what evolutionary forces might be acting on our brains right now. We know that the human brain has been shrinking over the many thousands of years since it reached its physical peak, but how might it change in future? The market for breast implants suggests that sexual selection is, let's just say, not yet entirely driven by how attractive a potential partner's brain is, so there is little evolutionary pressure for brain improvements. Yet many aspects of the environment in which our brains grow, work and adapt are changing rapidly, including the screens that have come to dominate many of our working lives as well as our free time. We can therefore make some predictions not only about what science and technology will make future brains better at, but also what our future brains may no longer need.

Smarter drugs

The obvious place to start is with smart drugs. Sometimes called nootropics, these are not a single category of drug, but rather a label given to any prescription medicine, over-the-counter remedy or even untested herbal supplement that is thought to improve cognitive function. The 2011 hit movie *Limitless* was a fantastic exploration of the possibilities and challenges that a really effective nootropic might bring to our world. A fictional nootropic, NZT-48, allows Bradley Cooper's character to unlock enormous unused capacity in his brain, transforming him from a struggling author to a man who earns millions on the stock market and stands for the US Senate. But along with this material success, the drug brings risk: many of those who take it die or are hospitalised due to its side-effects, and to maintain his supply the protagonist has to pursue a course that most would find ethically questionable. The film doesn't tell whether NZT-48 affects his long-term health, but it appears to make his life considerably more fulfilling, and certainly riskier. Crucially for the plot, the drug is illicit, secret and available only to a select few. This gives Cooper a major competitive advantage in whatever he chooses to do, but logistical problems in securing a safe ongoing supply of the drug.

Limitless is a great story. How close is it to reality? It certainly illustrates many of the real ethical questions involved in the development of cognitively enhancing drugs, including what side-effects would be acceptable as a trade-off for improved brain function, and how the drugs should be regulated so that their overall effect on society is positive and fair. The part that remains firmly in the realm of science fiction is just how effective NZT-48 was. Right now there simply aren't any drugs that have anything close to the effect on cognitive function portrayed in *Limitless*.

What we do have is a growing understanding of what smart drugs might look like, and some hint of how they might work. There are, for example, medicines that are licensed for Alzheimer's disease, which improve memory and other cognitive functions in most patients for a year or two. These treatments act by boosting the neurotransmitter acetylcholine, compensating for one of the mechanisms that has gone awry in this disorder. In a healthy young brain, which doesn't lack acetylcholine, taking these drugs doesn't do much to boost cognitive function. As a result, these types of drug are unlikely to be the basis for any *Limitless*-like revolution.

One group of prescription medicines is being used in the hope of a cognitive competitive advantage, however, and those are drugs that are used to treat attention deficit hyperactivity disorder (ADHD). ADHD is a developmental brain disorder that can seriously affect a person's school and work performance, and their social and family relationships. One class of drugs that seems to help people with ADHD are amphetamine-like stimulants such as methylphenidate (often sold under the brand name Ritalin). In people with ADHD, these drugs work by making more of the neurotransmitters noradrenaline and dopamine available to neurons in the prefrontal cortex, pepping up activity in this and other parts of the brain that affect cognition and that are thought to be underactive in ADHD.

There is evidence to suggest that people with no ADHD symptoms can also get some cognitive benefit from taking these drugs. Controlled research studies, which typically test a small number of healthy volunteers when taking the drug on one day and an identical-looking placebo on another day, seem to show that even in people without ADHD, methylphenidate can give a small boost to memory, and in some people may improve other aspects of cognition too.

That all sounds great, you might think, but are there any downsides? Well, many stimulant drugs are addictive, and have wide-ranging effects in the body, such as increased blood pressure and heart rate, and reduced sleep and appetite. These side-effects can be a serious risk to health, especially when the drugs are used frequently and without medical supervision. Therefore, their popularity as cognitive enhancers has been somewhat overtaken by drugs that have fewer reported side-effects. Probably the most popular right now is a drug called modafinil, which shares some properties with stimulants;

indeed, it was originally licensed to help people suffering from sleep disorders to stay alert and awake. It acts on the brain in a more nuanced way, at the same chemical site in the brain as cocaine, but in a different manner, which means it has a much lower likelihood of becoming addictive. As such, and because few people report any side-effects from taking modafinil, it has a relatively relaxed legal status in most countries.

It is estimated that around 90 per cent of the modafinil that is sold legally is prescribed by doctors, but for uses other than the sleep disorders for which it has been officially tested and cleared by regulatory authorities such as the US Food and Drug Administration. Instead it is being used to treat symptoms of fatigue and sedation caused by diseases or other medications, likely by military and other government agencies to keep their troops alert during extended combat or when on missions, and by students and other folk who believe it gives them a cognitive boost and a competitive advantage. This last market for unprescribed (illegal) sales of modafinil and other prescription-only smart drugs is thought to be large: surveys of US and European college students report that somewhere around 10–20 per cent have used one in the past year.

Modafinil has been reported to have two different kinds of benefit that explain its popularity among people who don't have sleep-related cognitive impairments. The first is that it seems to increase the pleasure that people get from engaging in tasks that might not be that appealing, such as knuckling down and studying for an exam. The second is that people tend to perform better on some measures of higher-level cognitive functions, such as working memory, planning and the ability to inhibit inappropriate or impulsive responses. These effects are usually small. But even small improvements in cognition, alertness or the ability to persist with a relatively boring task could have big functional benefits to many different types of people: hard-partying students, workers doing repetitive but safety-critical tasks, even top-performing surgeons, air-traffic controllers and military commanders working in conditions where fatigue and cognitive impairment could have deadly consequences.

In both the Alzheimer's and ADHD examples, the drugs work by correcting a neurochemical imbalance: the patient has less-than-optimal chemical function in some aspect of the brain and the drug corrects that. For modafinil, much of the benefit that healthy people report from taking it may be to do with increased motivation or ability to concentrate, rather than an increase in intelligence. What's not clear is whether any drugs that currently exist can improve cognition in someone who is already functioning optimally – i.e. who has no deficient neurochemistry, and is not sleep-deprived, bored or tired. Indeed, most of us already use a cognitive enhancer to give us a boost when we're tired or lacking focus: caffeine. Like other stimulants, excessive use of caffeine can cause jitters, palpitations and problems sleeping, but unlike all these newer drugs it's cheap, legal and universally available in a range of delicious marketed forms. All in all, it's hard to see that the nootropics we have available are that much of a step up from caffeine in terms of efficacy. They are certainly still some way from *Limitless* territory.

The main reason for this is that developing a new drug is a very tricky business. The pill that you or I end up swallowing is a complex piece of chemical engineering: it must dissolve in our stomach, be absorbed into the bloodstream and end up with its active ingredients available at the desired location in the body, and all at a rate that matches the effect we want. If it's designed to take away a headache, the desired rate is as quickly as possible. But if the goal is to correct an imbalance in the availability of serotonin in a particular system in the brain, we want it to do this at a steady pace, so that the amount of serotonin-related signalling in the brain is stable in between doses of medication.

To get to the brain, a drug must also overcome a massive physical challenge: crossing the blood-brain barrier (BBB). The BBB is a highly selective membrane that surrounds the brain and stops most of what is circulating in the bloodstream from being passed to the brain itself. The BBB exists specifically to keep substances that may be harmful to the brain away from it, such as substances needed for other bodily functions, and any toxins that might find their way into the bloodstream. The downside of the BBB is that it seriously limits the type of chemical that can be used in brain-targeted drugs. In particular, larger and more complex molecules find it hard to pass through. So if we are serious about future ways to enhance brain function, we may need to start to look beyond drugs.

Beyond drugs: new ways to activate the brain

As well as the technical challenges we just discussed, there's an inherent inefficiency in swallowing a pill in order to have an effect on the brain. Are there no more direct means of influencing brain circuits?

In fact, there are a lot, and some are already considered best-practice medical care. Electroconvulsive therapy (ECT) is an effective treatment used routinely in severe cases of depression and some other psychiatric conditions where drug treatment isn't working for a patient. It involves putting two electrodes on the patient's skull and passing a current between them. Not surprisingly, it was historically considered a last resort because in the early days the procedure was not optimised and significant side-effects, such as memory loss, were common. Modern ECT is a more sophisticated procedure designed to minimise these side-effects and takes place under sedation. A single round of ECT is effective in about half of patients who have major depression that has not responded to other types of treatment: a success rate that, for a psychiatric treatment, is very good indeed.

ECT is thought to work by resetting the electric and chemical activity in the brain in the short term, and perhaps by encouraging new neural growth in the longer term. There are several other forms of externally applied stimulation that are designed to produce somewhat less drastic neural resets. These can be applied to a more focused brain area and are not painful, so can be used when the patient is awake. The most common electrical form is transcranial direct current stimulation (tDCS), which applies a constant low current to an area of the brain using two electrodes placed on the scalp. The current can be used to excite or inhibit the firing of neurons in the area targeted by the electrodes: whilst it appears to neither improve nor worsen cognitive function, the technique shows promise for helping in depression, stroke and some other brain disorders, and is considered safe enough when used in single doses. At the moment, tDCS devices have been approved for treating depression in Europe but not the USA, where the amount of evidence for the procedure is not yet considered conclusive.

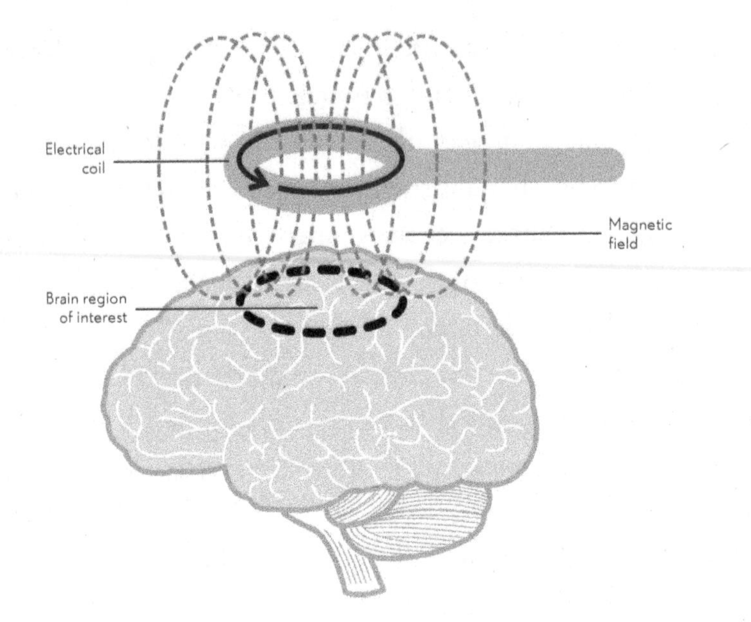

Figure 7: Transcranial magnetic stimulation (TMS)

TCDS is an important technique to watch, but others have been around longer, particularly one that uses magnets instead of electrical current to alter the firing of neurons. To give someone a dose of transcranial magnetic stimulation (TMS), you hold a specially designed electrical coil over their head, directing the angle onto the brain region of interest (see Figure 7). The coil then uses rapid changes in magnetic fields to induce a small electrical current in the brain. Although this may sound a little like a high-school physics lesson gone haywire, TMS is in fact fairly widely used and considered an effective treatment for disorders including migraine, neuropathic pain and depression.

There is, however, a catch. In the same way that drug therapies are limited by their need to cross through the brain's protection in the form of the blood-brain barrier, tDCS or TMS signals need to get through the skull in order to target the part of the brain where they can be useful. And the skull, being made of bone, is not a great conductor of electricity. As a result, these techniques have limitations in both how spatially accurate their stimulation can be and how far into the brain they can reach. For large areas on the external surface of the cortex that might not be a problem. But what if the bit you need to affect is small and deep inside the brain?

If that's the case then we move into the realms of direct brain stimulation, where electrodes are implanted in the brain, attached to wires fed under the skin and leading to a control device (positioned usually just below the collarbone). Clearly this surgery is not a step to be taken lightly but for patients with some types of Parkinson's disease, essential tremor, epilepsy, depression and obsessive-compulsive disorder, 'deep brain stimulation' (DBS) is considered a safe and effective treatment that can help symptoms where drugs cannot. Compared with the less invasive techniques like TMS and tDCS, implanting an electrode into exactly the right spot in the brain allows much more local and focused effects. The most successful of these have been implants in the basal ganglia, a group of very important nuclei located deep in the centre of the brain which control many aspects of voluntary movement as well as some aspects of cognitive function and emotion. When dopaminergic cells die in the basal ganglia, the many nasty symptoms of Parkinson's disease arise – including an inability to start certain movements and an inability to stop other parts of the body from moving. DBS has given patients back control over these disabling symptoms, and revolutionised the treatment of other neurological conditions where an inability to control one's own movements can be a huge barrier to everyday life.

Beyond biology: silicone implants for the brain

Almost without noticing it, we have moved to a world where we are talking about implanting machinery into the brain itself. DBS electrodes are one example of this, but they are just the start.

The idea that an electronic device can replace a damaged piece of neural hardware is not new: cochlear implants have been around since the 1980s. From the outside these devices look like a hearing aid, but what they do is quite different. Whilst a hearing aid is an external device that amplifies sounds going into the ear, a cochlear implant bypasses the ear to send electrical signals directly to the brain. It involves an array of electrodes being implanted into the cochlea, a chamber made of bone located in the inner ear. An external device sitting behind the ear picks up speech signals and converts them to electrical signals. These are picked up by the implanted array, which sends the signals directly to the auditory nerve. Cochlear implants were originally approved for adults who suffered hearing loss, but more recently there has been an emphasis on their use in young children where, because there is a critical period for language development, getting a cochlear implant before they reach eighteen months old is considered optimal. Many children who receive cochlear implants this early go on to develop close-to-normal language skills.

Analogous devices are now being worked on to provide vision for people who are blind. These might include an external video-capture system which processes dynamic real-time images and sends them as electrical signals to implanted electrode arrays. These could then relay them to the optical nerve, or directly to the primary visual cortex, the area where the earliest visual processing occurs.

As you can see, devices to compensate for sensory impairments are not just science fiction. But what other brain tools would any self-respecting cyborg want?

Bionic limbs, perhaps? Devices that bypass damaged or missing nerves to allow prosthetic limbs to be controlled directly by the brain are becoming increasingly sophisticated. These devices read neural activity from the motor cortex, decode the person's intent and then use the decoded signal to control a robotic limb, or external items such as a computer or a wheelchair.

One of the major limitations of such devices is their ability not just to read out signals from a brain but also to provide real-time sensory feedback, so that the user can fine-tune the control of the prosthetic. (Imagine how hard it would be to cut up meat without being able to feel how much pressure you are applying to the knife, or eat a banana without knowing how hard you are gripping it.) Much of the technological development in this area has been led by military-funded researchers, particularly in the US, where the large number of veterans who return from warzones with life-changing injuries to the brain or body provide a huge impetus to investing in these cutting-edge technologies. In 2015, the US defence-research agency DARPA announced that it had managed, for the first time, to close a feedback loop between a prosthetic hand and areas of the sensory and motor cortex. The recipient, a twenty-eight-year old man who had been paralysed by a spinal-cord injury for more than a decade, reported that he could not only control the hand directly from his brain, but also feel physical sensations in it, just as in his biological hand. In March 2017, the BBC and other news sources reported that a quadriplegic man named Bill Kochevar had been able to feed himself mashed potato by using sensors implanted in the motor cortex to control implants in his arm, allowing him to move it for the first time since a cycling accident eight years previously. Cases like this show how quickly we may be able to make major steps forward in the useful integration of minds and machines.

Another really useful type of brain–computer interface is one that can aid communication where damage to the speech-production system has occurred. Perhaps the most famous user of such systems is Professor Stephen Hawking, who has been living with the symptoms of motor neurone disease for over fifty years. As a professor at the Department of Applied Mathematics and Theoretical Physics in Cambridge, you'd expect him to be well placed to have access to the very best brain-controlled communication aids. Having experimented with many assistive systems, including brain- and eye-movement-controlled interfaces, Professor Hawking reports that the system he finds most convenient and least tiring is in fact relatively simple. It involves an infrared switch attached to the frame of his glasses, which detects when he twitches his cheek and uses this to control the movement of a cursor on his computer. The text he writes on the computer is then sent to a speech synthesiser, allowing him to take part in real-time conversations as well as pre-recorded lectures and speeches.

Professor Hawking's communication system has allowed him to pursue a phenomenally successful career despite the ravages of disease in the motor areas of his brain. Compensating for the effects of brain disease and brain injury using external technologies looks likely to be a huge step forward for some patients. It also seems very likely that brain–machine interfaces and neural prosthetics will go on to develop in ways that become useful to healthy brains, perhaps taking us past our current limits in speed, strength, information-processing or memory. Artificial intelligence (AI) systems, which harness the use of enormous external computing power to solve problems, are now starting to perform better than human intelligence in some specific fields. As yet these are nowhere near as flexible as the human brain – not a surprise, since they cannot match it for complexity, number of parts or the many millennia over which it has been optimised. But AI can do some specific processing faster than a human, and with more tenacity, no fatigue and no human error. For example, AI systems have repeatedly beaten the top living humans at games like chess and Go, where the rules of the game are fixed and finite, and the number of possible permutations and outcomes of a move can be calculated more efficiently by an AI system than by the limited and fallible human brain. As prosthetic and brain–machine interface technologies mature, combining them with AI-powered processing systems may help us on our way to cyborg levels of function.

In the meantime, for patients with brain disorders, is silicon technology the only hope? Actually, for some patients, there is reason to believe that the biggest progress may come not from better computers or machines but from biological progress, which may allow us to correct the pathology of disease as close as possible to the source of the problem.

Making biology better, at the source

As we have discussed already, most brain disorders and brain functions have complex genetic origins, with not one but many genes playing a role, and many environmental factors too. Whilst it is likely that sooner or later babies will start to have their genomes sequenced at birth, this complex genetic landscape, plus the accompanying convolution of epigenetics, means that we are unlikely to be able to predict much about their future personality, IQ or risk for disorders based only on that sequence.

Genomics are nonetheless likely to play a much bigger role in treating brain disorders in the future, because gene therapies can alter a genetically driven trait or biological function at source. The basic idea here is that genetic material would be used like a drug, to correct or compensate for faulty biology, but in a way that could be targeted to a specific organ or group of cells, and potentially turned on and off as needed.

Let's return for a moment to the basal ganglia. We discussed before how the death of dopaminergic cells here leads to Parkinson's disease. Another way that this crucial bit of the brain can go wrong is if its neurons are exposed to a mutated form of the huntingtin protein. As you might guess from the name, mutations in the gene that codes for the huntingtin protein are the cause of Huntington's disease, a rare example of a brain disease caused by a single gene. Stopping the symptoms of Huntington's disease by stopping the ganglia neurons from dying would therefore be equally simple – if only we could stop the mutant huntingtin protein from forming.

The first trials of a new genetic therapy that aims to do just that began in late 2015. The therapy itself is a small molecule called IONIS-HTTRx, which is injected into the spine and travels up in the cerebrospinal fluid, eventually reaching neurons in the brain. The hope is that the drug will silence the huntingtin gene in the cells it reaches, greatly reducing the amount of huntingtin protein that is produced in them. Gene-silencing works by targeting not the DNA itself but rather RNA, the intermediate chemical that is needed to translate the recipe that's written in the DNA into the long chain of amino acids that forms a particular kind of protein. So the DNA remains intact but the RNA is inactivated, stopping the cell from making the nasty kind of the huntingtin protein.

It's too early to tell whether this particular therapy, or ones that work in similar ways, will be safe and effective in humans. But gene-silencing isn't the only trick that's heading for a brain near you. Recently there have been amazing advances in techniques that allow the actual DNA code to be edited, removing the mutated sequence altogether. These techniques are known by complicated names like clustered regularly interspaced short palindromic repeats (CRISPR) and zinc finger nucleases (ZFNs). They rely on tricks that allow a particular sequence of DNA to be recognised and the DNA molecule cut at that point.

What this means in practice is that if we could safely get some of these editing tools into a living cell, they could be programmed to cut out just the disease-causing piece of the DNA sequence, and replace it with a neutral piece – like using a word processor to censor an email before it is sent. The 'if' in that sentence is important: these gene-editing tools are large molecules that cannot be given as a drug but must be injected directly into the brain, or packaged up in a specially designed virus and then passed to the brain through a surgical procedure. In an early-stage embryo, it may be relatively easy to get such a virus in to target all cells, from which a Huntington's disease-free brain could then grow. In an adult brain it is going to be a whole lot harder to ensure that every neuron that could be affected by the disease is infected with the beneficial virus.

Huntington's disease is unusual in being caused by a single gene, but it is not the only brain disorder where advances in genetic techniques may prove to be the step-change that is so urgently needed. In Parkinson's disease, for example, there are currently several virus-based approaches that have shown some promise in early clinical trials. These aim to deliver gene therapies that can either encourage the growth of neurons in specific areas, or increase or reduce the production of neurochemicals in particular parts of the basal ganglia.

One final exciting possibility is the use of genetic engineering to create a system where different groups of neurons in the brain could be turned on or off, simply by shining a light on them. This technique, known as optogenetics, sounds far-fetched but is already widely used in laboratory research. It relies on the fact that neurons and other cells can be genetically engineered, in much the same ways as we discussed earlier, to express light-sensitive ion channels. Ion channels are proteins that sit in the cell wall; they act as a gate that lets in or keeps out different kinds of ions (charged particles). In neurons, ion channels are particularly important because they determine when a neuron will fire, by controlling the flow of charged particles and hence the electrical state of the cell. Once a cell has these new light-sensitive ion channels, the firing of that neuron (or group of neurons) can be controlled by literally flashing a light on to the cell. The technique allows spectacularly specific manipulation of neural activity: not only can researchers control which neurons are made sensitive to the light, they can control precisely when the firing in these cells is turned on and off.

It's a lovely technique which is proving a really useful way of probing the behaviour of different neural circuits in lab experiments in living animals. You may have spotted the hitch, though: it still relies on that old problem of getting a bit of engineered genetic code into the brain – and now you also need to be able to get a light source in there too! So is there really any hope that this will result in a therapy that can help patients any time soon?

Remarkably, some brand-new research from the Massachusetts Institute of Technology suggests that this might be closer than we think. The researchers, led by an extraordinary scientist named Li-Huei Tsai, took as their starting point the knowledge that in patients with Alzheimer's disease, one of the many things that goes wrong in the brain is a reduction in gamma waves. These are background brainwaves (more formally, 'neural oscillations') of a particular frequency, around forty per second (40 Hz). Gamma waves are interesting because they are produced by groups of neurons that seem to play a role in cognitive processing, which, as we know, is impaired early in Alzheimer's disease. Tsai's own 'brainwave' was this: what if we could pump up the gamma waves in someone with Alzheimer's disease? Would that then restore cognitive function? The way to do this, she thought, was optogenetics.

The team took some mice that had been genetically engineered as a model of Alzheimer's disease, which showed the characteristic amyloid brain plaques and had problems with learning and memory. They then infected their brains with a virus carrying a light-sensitive ion channel, and then drilled a tiny hole in the skull so that an optic fibre could be inserted. This allowed them to activate neurons by flashing light at a selected speed – yep, you guessed it: 40 Hz. They hoped that by artificially boosting gamma waves in the brain, they might eventually be able to improve the mice's Alzheimer-like symptoms. In fact, what they found was astonishing: after just one hour of the light treatment, the amount of amyloid in the brain was reduced by half.

So artificially inducing gamma waves in mice seemed to be removing some of what we think may be the basic biological cause of Alzheimer's. How on earth could they find out if the same would be true in humans? It would be pretty hard to get anyone to agree to having holes drilled into the skull so that the light could be flashed in.

Luckily, nature has already provided us with a way to get light into the skull: the eye sockets. Tsai's group went on to demonstrate that simply putting the mice in a room with lights that flashed at 40 Hz was just as effective in clearing amyloid from the brain. Even more astounding, inducing gamma waves by flashing the light at a specific set of neurons seemed to reactivate a memory that an Alzheimer-like mouse had forgotten. So the technique may offer hope not only in stopping or reversing the build-up of the disease in the brain, but also in recovering memories that have already been lost. They are hoping to start trials in humans soon. We'll be watching this space with great interest!

The evolving brain

With so many amazing scientific and technological advances, it's easy to lose sight of changes that may be happening organically right now. Throughout our exploration of brain function, we have often referred to evolution, and it's worth bearing in mind that this process is not over.

How we spend our time, and hence the uses we put our brains to, continues to change. Indeed, it has changed especially rapidly in the developed world over the past century. For more and more people, the ability to hold down gainful employment depends not only on physical parameters but on the adequate function of cognitive processes. A hundred years ago, most people had manual jobs that could not be done with, say, a broken leg, but that were reasonably kind to people with a poor memory or a lack of planning skills. Now the reverse is true, and many of us spend eight or more hours a day exercising our minds, and very little time exercising our bodies.

In the past ten years or so, smartphones and tablets have become a ubiquitous feature of many people's lives. These devices have decreased the need for some aspects of brain function, for example remembering phone numbers and how to get to distant locations. They have increased the need for other skills, such as fine motor control: witness the physical dexterity displayed by a teenager texting with one hand, or a child gaming on a handheld device. Above all, the availability of computers, smartphones, tablets, e-readers and plasma TVs has increased the amount of time the average person in the developed world spends staring at a backlit screen.

We're constantly exposed to media reports asserting that we should be freaking out about the potential harms of screen time. Does it really matter? The simple answer is we don't know what effect this has on our brains and, especially important in parents' minds, the brains of developing children.

There are certainly reasons why we should consider limiting screen time. For one thing, time spent in front of a screen is largely physically inactive and we know that physical activity is important for healthy development (indeed, for maintaining good health at all ages). However, the limited evidence available at the moment suggests that screen time has no effect on the amount of outdoor play that young people take part in. For another, it's plausible that exposure to bright screens late at night disrupts circadian rhythms, making it difficult to get enough good-quality sleep. One interesting Swiss study had teenagers use blue-light-blocking glasses for a week when using computers in the evening. They found that the glasses reduced both the impact of screen time on melatonin (measured in their saliva) and late-night alertness. It didn't however seem to have any effect on sleep quality or functioning the morning after, suggesting that there may not be an immediate reason to panic about the effects of late-night screen use on teenagers' lives.

What's not at all clear is whether what children do while staring at those screens is good or bad for the brain. We simply don't know whether spending hours gaming is good for cognitive skills like mental rotation, and increasing social interactions among shy kids, or conversely whether it reduces children's attention spans and makes them less able to engage with the offline world. Whilst it is natural for the press to prey on parents' fears, the weight of opinion among developmental psychologists and other academic experts is that at present we simply don't have much evidence either way.

Aside from how we use them, our brains may also be affected by other aspects of modern life. As we noted in earlier chapters, making healthy choices in diet, physical activity, and drug, alcohol and tobacco use affects the brain in much the same way as it does other parts of the body. Modern environmental legislation and employment law, among other things, protect us from many of the toxic exposures that might historically have been acceptable. (It is, after all, only a decade since every pub or bar-worker in the country spent most of their working hours in an impenetrable smoky haze.) Exposure to lead from petrol, to mercury in paints, asbestos in insulation and high levels of pesticides in the human food chain, have all decreased enormously over the past few decades. Public-health messages about the importance of things like using sunscreen and not drinking alcohol or smoking during pregnancy mean that children today are probably exposed to far fewer neurotoxins and mutation-causing carcinogens than at any time since the Industrial Revolution.

We expect to live considerably longer today, and so what we need now are brains that are more resilient to the ravages of old age and later-life diseases such as Alzheimer's. Since these strike after the age of reproduction, natural selection will not be a factor in brain evolution. But studying diseases like Alzheimer's can give us an insight into which modern behaviours and environments influence brain resilience throughout life, and the mechanisms by which that might occur. For example, a recent study following the lives of more than 6 million Canadians found that living close to a busy road increased the risk for Alzheimer's disease, presumably through the effects on the brain of either noise or air pollution. Identifying such top-level risk factors can help us understand more about the basic processes in the brain: for example, whilst we have known for a while that sleep is important in Alzheimer's, only recently have we discovered one way in which this works. It turns out that one of the mechanisms of homeostasis, which maintains the brain in a healthy state, is that during sleep the brain's interstitial space expands by 60 per cent. This massively increases the exchange of waste products between the interstitial fluid, which surrounds the neurons themselves, and the cerebrospinal fluid, which surrounds the brain and runs down through the spinal cord. This allows the brain to send neurotoxins, including that old foe amyloid, out of the brain via the cerebrospinal fluid, helping keep neurons immersed in a healthy bath of nutrients rather than toxins.

Will future brains be better or worse than current ones?

As you'll have come to understand, there is a huge amount we don't yet know about the workings of the brain. But we can nonetheless speculate a little about how current trends in the way we use our brains, together with changes in our broader environment, may come to be reflected in the brain itself.

The brain is an organ that shapes itself in accordance with the environment it develops in. When we think about the strengths and weaknesses of future brains we are mostly considering how early experiences will shape the nature of each individual's brain, rather than slow-moving evolutionary processes. Few of the changes we note here will affect reproductive success; indeed, since increased education and income tend to reduce the number of children a woman has, it may be that the more cognitively optimal a brain becomes, the smaller number of offspring will inherit its genes. Changing times are likely to favour brains with different strengths from those that have been selected for so far in human history.

So what might we expect future brains to become better at? As discussed earlier, since we now use our large muscle groups less and our fine motor skills more, we might expect some redistribution of space given over to areas of movement control in the motor cortex, and a modified cerebellum. If robots and other forms of AI take over many of the lower-skilled and lower-paid jobs in hospitality, retail, factories and service industries, will this free up more people and leisure time to take part in creative endeavours? If so, this may strengthen the connectivity in the 'default mode network', a network of interacting brain areas that is particularly active during mind-wandering and imaginative tasks. And as pharmaceuticals, genetic therapies and neural prosthetics become more effective, will we choose to allow healthy people to benefit from them, perhaps opening up new sensory realms where we can see ultraviolet wavelengths or sense magnetic fields? The answers to these questions depend heavily on what societies choose: the nature of our future brains will be determined by legal and policy advances as well as technological ones.

There is one way in which we are actively tipping the balance in the physical evolution of the brain. We started our discussion of the limits of the human brain by pointing out that head size at birth determines how big and how well developed a baby's brain can be. Advances in obstetric medicine, and particularly the rapid rise of the Caesarean section as a way of safely delivering a large baby, means that this key evolutionary pressure may now be irrelevant for most women in the developed world. With around a quarter of babies delivered by C-section in countries like the UK and USA, it is estimated that this has already led to a 10–20 per cent increase in the number of large babies safely born to women with narrow pelvises (though large babies do not necessarily make for better brains).

Indeed, there are some brain functions that you might well expect to be weaker in future generations. In the same way that our mobile phones have removed the need for us to remember people's contact details, technology will continue to replace the need for skills traditionally acquired through diligence and practice. Driverless cars, for example, will replace not only the need to learn the motor controls necessary to balance accelerator, clutch and brake, but will also reduce the hours of concentration needed to travel on busy roads. Will new tasks requiring equally hard-won skills come to replace the driving test as a rite of passage to adulthood? Or will today's babies simply have more limited attention spans because they are never required to practise anything as arduous as the long commute home on a busy motorway?

What do our experts think the future holds for the brain?

When we interviewed the experts featured in this book we asked them all the following question: How will the human brain evolve? It was a toughie, that was for sure, but despite squirming slightly at having to give an answer, they all proposed thoughtful ideas of what we might expect to see. Furthermore, even though they come from different fields and so have different perspectives on the brain, they all independently identified new technologies as playing a major role in how our brain may change over time.

Dr Graham Murray: 'The brain won't evolve particularly in the near future. It will take a long time for significant evolutionary changes to occur. However, we can already modulate its structure and function.

'In the 1940s and 50s an operation called lobotomy was conducted on many thousands of patients to try and help with psychiatric symptoms. This involved the removal of significant parts of the frontal lobe of the brain. Although sometimes there was an improvement in some symptoms, the side-effects were often catastrophic, and thankfully the process was abandoned. In psychiatric practice nowadays, the main way we try to improve brain function is with drugs. However, it is possible that with advances in the scientific understanding of the brain we may in future be able to offer new kinds of therapies to help psychiatric patients. As an example, deep brain stimulation is already quite commonly used in Parkinson's disease and has shown promise in small trials of patients with obsessive-compulsive disorder. This kind of new approach is likely to increasingly become an alternative to drug treatments for people with severe psychiatric problems. The key will be to conduct rigorous randomised controlled trials to prove that these therapies are safe and effective, so that we don't repeat the mistakes of the lobotomy era.

'And it may not stop there. It may be the case that we start seeing the increasing use of interventions to enhance the abilities of healthy brains in the general population. It will be fascinating to see what the future brings.'

Dr Lauren Weiss: 'One thing that I think might have a big effect is the rise of gene-editing technologies, which are just starting to come into play. These could play a really big role in allowing us to prevent specific brain conditions, perhaps in combination with prenatal screening programmes.'

Dr Simon Kyle: 'I think that technology is going to play a major role – the rapid increase in use of technology over the last twenty years does seem to be having an effect on our cognitive and social processes, and I imagine that will influence how the brain will evolve and how it will appear in the future. Ultimately I think the major evolution will be in how we approach brain function and the development of personalised techniques to directly interfere and optimise brain processes through technological and genetic engineering. The quest for cognitive enhancement and efficiency will be at the forefront. Here it is worth remembering that good-quality, restorative sleep may be one of the best cognitive enhancers we have!'

Dr Fergus Gracey: 'I think it's really about how society will evolve. For example, as technology evolves and is at our fingertips how will our brain evolve in response? Will "the market" actually influence how our brains evolve as it produces, and convinces us to buy, more advanced products? Will we become more influenced by marketing or will we be better able to resist it? What does this mean for our cognition and how we understand and measure intellect? For example, will the cognitive assessments we currently use even be valid in twenty years' time?'

Ms Maggie Alexander: 'In each era, as new opportunities, new technologies and new circumstances arise, we lose some brain functions but gain others. For example, losing map-reading skills but gaining the ability to use a sat nav; it's a different set of skills but still very useful. The pace of change over the last 100 years has not been linear; it has almost been logarithmic. So whilst we can't predict what may come next, we will adapt;

of that I have no doubt. Despite this, you will always have half a generation of people who feel left behind and who struggle more than others to adapt. We're on the cusp of the driverless car, for instance; something I don't like the thought of but I can see will be a huge liberation for others. Think of the many people who currently have difficulty in getting around and need assistance, such as those with physical or visual impairments, who may one day be able to benefit from such a technology. This may in turn lead to beneficial adaptations in the brain.'

How much brain will future humans need?

With such potential for technology to support functions currently reliant on fallible brain processes, an optimist might conclude that future brains could function well with less capacity, or with greater amounts of damage.

A pessimist might reply that that depends entirely on societal choices: will we be glad of a diverse world where people live with more age-induced brain degeneration for longer, where those at the lower ends of the genetic IQ lottery can find no employment that a robot cannot do better, or where increased population pressure means that some of our recent gains in IQ are reversed by the effects of insufficient nutrition and an increasingly polluted environment?

The optimist might counter that these issues too may be solvable with technology, or that what is necessary for a minimally viable brain might change considerably as mankind's future unfolds. If we go into space, for example, the need for societal harmony during long space voyages and in small pioneer communities might make social cognition the most highly prized skill.

There may already be aspects of brain function that we would be glad to lose, where there is a mismatch between our brain's evolutionary past and the world we now live in. For example, many of the one in four people who suffer symptoms of anxiety and depression each year might welcome a reduction in function of their limbic system, which processes emotion, their autonomic nervous system's fight-or-flight responses; or the hypothalamic–pituitary–adrenal axis that moderates responses to stress. Such systems, which during evolution helped us to stay alive, may be actively unhelpful in the overwhelming majority of non-life-threatening contexts in which we usually find ourselves. In a world of twenty-four-hour news, a neurally driven stress response to a shooting that happened many thousands of miles away may help keep you engaged with a news site, but it does nothing for your chances of survival, or of a good night's sleep.

If we've learned anything on our tour through past, present and future brains, it's that humans have, and seem likely to retain, a great capacity to function in the most extraordinarily diverse circumstances. In each of us, the fickle hands of disease, damage and genetic and environmental fortune set us up with a brain that is predisposed to certain tendencies. Those tendencies play out across our lifetime in complex and interacting ways that will certainly take more than our lifetimes to untangle. We hope you have enjoyed the ride so far and appreciate the brain you currently have, however much of it there may be!

Bibliography

Chapter 1

Anderson, B., and Harvey, T., 'Alterations in Cortical Thickness and Neuronal Density in the Frontal Cortex of Albert Einstein', *Neuroscience Letters*, June 1996

Australian Museum, 'How Have We Changed Since our Species First Appeared?', http://australianmuseum.net.au/how-have-we-changed-since-our-species-first-appeared, October 2015

Benson-Amram, Sarah, Dantzer, Ben, Stricker, Gregory, et al., 'Brain Size Predicts Problem-solving Ability in Mammalian Carnivores', *Proceedings of the National Academy of Sciences of the United States of America*, March 2016

Bohn, Lauren E., 'Q&A: "Lucy" Discoverer Donald C. Johanson', *Time*, March 2004

Bozek, Katarzyna, Wei, Yuning, Yan, Zheng, et al., 'Exceptional Evolutionary Divergence of Human Muscle and Brain Metabolomes Parallels Human Cognitive and Physical Uniqueness', *PLoS Biology*, May 2014

Brunet, Michel, Guy, Franck, Pilbeam, David, et al., 'A New Hominid from the Upper Miocene of Chad, Central Africa', *Nature*, July 2002

Cadsby, Ted, *Closing the Mind Gap: Making Smarter Decisions in a Hypercomplex World*, Toronto: BPS Books, 2014

Cairó, Osvaldo, 'External Measures of Cognition', *Frontiers in Human Neuroscience*, October 2011

Carmody, R. N., Weintraub, G. S., and Wrangham, R. W., 'Energetic Consequences of Thermal and Nonthermal Food Processing', *Proceedings of the National Academy of Sciences of the United States of America*, November 2011

Clark, W. E. Le Gros, *The Fossil Evidence for Human Evolution*, University of Chicago Press, 1955

Cosgrove, K. P., Mazure, C. M., and Staley, J. K., 'Evolving Knowledge of Sex Differences in Brain Structure, Function and Chemistry', *Biological Psychiatry*, October 2007

DeFelipe, Javier, 'The Evolution of the Brain, the Human Nature of Cortical Circuits, and Intellectual Creativity', *Frontiers in Neuroscience*, May 2011

Douglas Fields, R., 'Change in the Brain's White Matter', *Science*, November 2010

Errington, Jeff, 'L-form Bacteria, Cell Walls and the Origins of Life', *Open Biology*, Royal Society Publishing, January 2013

Gould, Stephen Jay, *Ever Since Darwin: Reflections in Natural History*, London: Penguin, 1991

Gunz,. P., Neubauer, S., Maureille, B., et al., 'Brain Development after Birth Differs Between Neanderthals and Modern Humans', *Current Biology*, November 2010

Hawks, John, 'How Has the Human Brain Evolved?', *Scientific American*, July 2013

Herculano-Houzel, Suzana, 'The Remarkable, Yet Not Extraordinary, Human Brain as a Scaled-Up Primate Brain and Its Associated Cost', in Striedter, G. F., Avise, J. C., and Ayala, F. J., eds., *In the Light of Evolution: Volume VI: Brain and Behavior*, National Academies Press, 2013

Herculano-Houzel, Suzana, and Kaas, John H., 'Gorilla and Orangutan Brains Conform to the Primate Cellular Scaling Rules: Implications for Human Evolution', *Brain, Behavior and Evolution*, February 2011

Hofman, Michel A., 'Evolution of the Human Brain: When Bigger is Better', *Frontiers in Neuroanatomy*, March 2014

Institute of Human Origins, 'Homo Erectus', http://www.becominghuman.org/node/homo-erectus-0, 2008,

Kappelman, John, 'The Evolution of Body Mass and Relative Brain Size in Fossil Hominids', *Journal of Human Evolution*, March 1996

Liu, C., Tang, Y., Ge, H., Wang, F., Sung, H., et al., 'Increasing Breadth of the Frontal Lobe but Decreasing Height of the Human Brain between Two Chinese Samples from a Neolithic Site and from Living Humans', *American Journal of Physical Anthropology*, May 2014

McAuliffe, Kathleen, 'If Modern Humans Are So Smart, Why Are Our Brains Shrinking?', *Discover*, September 2010

Oró , J. J., 'Evolution of the Brain: From Behavior to Consciousness in 3.4 Billion Years', *Neurosurgery*, June 2004

Rakic, Pasko, 'Evolution of the Neocortex: A Perspective from Developmental Biology', *Nature*, October 2009

Robson, David, 'A Brief History of the Brain', *New Scientist*, September 2011

Rosenberg, Karen, and Trevathan, Wenda, 'Birth, Obstetrics and Human Evolution', *BJOG*, November 2002

Smithsonian National Museum of Natural History, 'Sahelanthropus Tchadensis', http://humanorigins.si.edu/evidence/human-fossils/species/sahelanthropus-tchadensis

Ibid., 'Bigger Brains: Complex Brains for a Complex World', http://humanorigins.si.edu/human-characteristics/brains, February 2016

Stringer, Christopher, 'Why Have Our Brains Started to Shrink?', *Scientific American*, November 2014

UCL News, 'Human Evolution Driven by Climate Change', https://www.ucl.ac.uk/news/news-articles/1310/171013-Human-evolution-driven-by-climate-change, October 2013

Ventura-Antunes, Lissa, Mota, Bruno, and Herculano-Houzel, Suzana, 'Different Scaling of White Matter Volume, Cortical Connectivity, and Gyrification across Rodent and Primate Brains', *Frontiers in Neuroanatomy*, April 2013

Wayman, Erin, 'Why Are Humans Primates?', Smithsonian.com, October 2012

Webb, Jeremy, 'Richard Wrangham: Cooking Is What Made Us Human', *New Scientist*, December 2009

Wildman, Derek E., Uddin, Monica, Liu, Guozhen, et al., 'Implications of Natural Selection in Shaping 99.4% Nonsynonymous DNA Identity between Humans and Chimpanzees: Enlarging Genus *Homo*', *Proceedings of the National Academy of Sciences of the United States of America*, April 2003

Wood, Bernard, 'Human Evolution: Fifty Years after Homo Habilis', *Nature*, April 2014

World Health Organization, 'Dementia', http://www.who.int/mediacentre/factsheets/fs362/en/, April 2016

Chapter 2

Barton, J. J., Press, D. Z., Keenan, J. P., and O'Connor, M., 'Lesions of the Fusiform Face Area Impair Perception of Facial Configuration in Prosopagnosia', *Neurology*, January 2002

Callaway, Ewen, 'Starvation in Pregnant Mice Marks Offspring DNA', *Nature*, July 2014

Clutton-Brock, Tim, 'Cooperation Between Non-kin in Animal Societies', *Nature*, November 2009

Crick, F. C., and Koch, C., 'What is the Function of the Claustrum?', *Philosophical Transactions of the Royal Society of London*, June 2005

Darwin, Charles, *The Descent of Man and Selection in Relation to Sex*, New York: D. Appleton & Co., 1871, 1896, p. 66

Davidson, R. J., 'One of a Kind: The Neurobiology of Individuality', *Cerebrum*, June 2014

Dunn, Rob, 'Your Appendix Could Save Your Life', *Scientific American*, January 2012

Human Intelligence, 'Charles Darwin', http://www.intelltheory.com/darwin.shtml, December 2016

Jarvis, Erin, 'Humans – Are We Just Another Primate?', *Berkeley Science Review*, August 2011

Eunice Kennedy Shriver National Institute of Child Health and Human Development, 'What Role Do Epigenetics and Developmental Epigenetics Play in Health and Disease?', https://www.nichd.nih.gov/health/topics/epigenetics/conditioninfo/Pages/impact.aspx

Leech, R., and Sharp, D. J., 'The Role of the Posterior Cingulate Cortex in Cognition and Disease', *Brain*, January 2014

Minagar, A., Ragheb, J., and Kelley, R. E., 'The Edwin Smith Surgical Papyrus: Description and Analysis of the Earliest Case of Aphasia', *Journal of Medical Biography*, May 2003

Penfield, Wilder, and Boldrev, Edwin, 'Somatic Motor and Sensory Representation in the Cerebral Cortex of Man as Studied by Electrical Stimulation, *Brain*, 1937

Randal Bollinger, R., Barbas, A. S., Bush, E. L., Lin, S. S., and Parker, W., 'Biofilms in the Large Bowel Suggest an Apparent Function of the Human Vermiform Appendix', *Journal of Theoretical Biology*, December 2007

Roth, G., and Dicke, U., 'Evolution of the Brain and Intelligence in Primates', *Progress in Brain Research*, 2012

Schlaug, Gottfried, Norton, Andrea, Overy, Katie, and Winner, Ellen, 'Effects of Music Training on the Child's Brain and Cognitive Development', *Annals of the New York Academy of Sciences*, 2005

Smith, Kerri, 'Evolution of a Single Gene Linked to Language', *Nature*, November 2009

Thomson, Helen, 'Famine Puts Next Two Generations at Risk of Obesity', *New Scientist*, July 2014

Wickens, Andrew P., *A History of the Brain: From Stone Age Surgery to Modern Neuroscience*, Psychology Press, 2014, p. 9

Zahid, A., 'The Vermiform Appendix: Not a Useless Organ', *Journal of the College of Physicians and Surgeons Pakistan*, April 2004

Zaidel, Dahlia W., 'Creativity, Brain, and Art: Biological and Neurological Considerations, *Frontiers in Human Neuroscience*, June 2014

Chapter 3

Alzheimer's Association, '2014 Alzheimer's Disease Facts and Figures', *Science Direct*, March 2014

Angold, A., Costello, E. J., and Erkanli, A., 'Comorbidity', *Journal of Child Psychology and Psychiatry*, January 1999

Auyeung, B., Baron-Cohen, S., Ashwin, E., Knickmeyer, R., Taylor, K., Hackett, G. and Hines, M., 'Fetal Testosterone Predicts Sexually Differentiated Childhood Behavior in Girls and in Boys', *Psychological Science*, February 2009

Baron-Cohen, S., 'The Extreme Male Brain Theory of Autism', *Trends in Cognitive Sciences*, June 2002

Chyi, L. J., Lee, H. C., Hintz, S. R., Gould, J. B., and Sutcliffe, T. L., 'School Outcomes of Late Preterm Infants: Special Needs and Challenges for Infants Born at 32 to 36 Weeks Gestation', *Journal of Pediatrics*, July 2008

Cohen, Jacob, *Statistical Power Analysis for the Behavioral Sciences*, Academic Press, 1969

Costa, P. T. Jr, Terracciano, A., and McCrae, R. R., 'Gender Differences in Personality Traits across Cultures: Robust and Surprising Findings', *Journal of Personality and Social Psychology*, August 2001

Crews, F. T., and Boettiger, C. A., 'Impulsivity, Frontal Lobes and Risk for Addiction', *Pharmacology, Biochemistry, and Behavior*, September 2009

Faul, Mark, Xu, Likang, Wald, Marlena M., and Coronado, Victor G., 'Traumatic Brain Injury in the United States: Emergency Department Visits, Hospitalizations, and Deaths 2002–2006', US Department of Health and Human Services, March 2010

Feingold, A., 'Gender Differences in Personality: A Meta-analysis', *Psychological Bulletin*, November 1994

Fombonne, E., 'Epidemiological Surveys of Autism and Other Pervasive Developmental Disorders: An Update', *Journal of Autism and Developmental Disorders*, August 2003

Ibid., 'Epidemiology of Pervasive Developmental Disorders', *Pediatric Research*, June 2009

Gogtay, N., Lu, A., Leow, A. D., Klunder, A. D., Lee, A. D., et al., 'Three-dimensional Brain Growth Abnormalities in Childhood-onset Schizophrenia Visualized by Using Tensor-based Morphometry', *Proceedings of the National Academy of Sciences of the United States of America*, October 2008

Hanamsagar, Richa, 'Sex Differences in Neurodevelopmental and Neurodegenerative Disorders: A Largely Ignored Aspect of Research', *Current Neurobiology*, January 2015

Horwath, E., and Weissman, M. M., 'The Epidemiology and Cross-national Presentation of Obsessive-Compulsive Disorder', *Psychiatric Clinics of North America*, September 2000

Hyde, J. S., 'Gender Similarities and Differences', *Annual Review of Psychology*, 2014

Ibid., 'Sex and Cognition: Gender and Cognitive Functions', *Current Opinion in Neurobiology*, June 2016

Jacquemont, S., Coe, B. P., Hersch, M., Duyzend, M. H., Krumm, N., et al., 'A Higher Mutational Burden in Females Supports a "Female Protective Model" in Neurodevelopmental Disorders', *American Journal of Human Genetics*, March 2014

Jeronimus, B. F., Kotov, R., Riese, H., and Ormel, J., 'Neuroticism's Prospective Association with Mental Disorders Halves after Adjustment for Baseline Symptoms and Psychiatric History, but the Adjusted Association Hardly Decays with Time', *Psychological Medicine*, October 2016

Kessler, R. C., McGonagle, K. A., Swartz, M., Blazer, D. G., and Nelson, C. B., 'Sex and Depression in the National Comorbidity Survey I: Lifetime Prevalence, Chronicity and Recurrence', *Journal of Affective Disorders*, October–November 1993

Kessler, R. C., Sonnega, A., Bromet, E., Hughes, M., and Nelson, C. B., Posttraumatic Stress Disorder in the National Comorbidity Survey', *Archives of General Psychiatry*, December 1995

Kim, Y. S., Leventhal, B. L., Koh, Y. J., Fombonne, E., Laska, E., et al., 'Prevalence of Autism Spectrum Disorders in a Total Population Sample', *American Journal of Psychiatry*, September 2011

Lai, Meng-Chuan, Lombardo, Michael V., Auyeung, Bonnie, Chakrabarti, Bhismadev, and Baron-Cohen, Simon, 'Sex/Gender Differences and Autism: Setting the Scene for Future Research', *Journal of the American Academy of Child and Adolescent Psychiatry*, January 2015

Manzardo, A. M., Madarasz, W. V., Penick, E. C., Knop, J., Mortensen, E. L., et al., 'Effects of Premature Birth on the Risk for Alcoholism Appear to Be Greater in Males than Females', *Journal of Studies on Alcohol and Drugs*, May 2011

Moore, David S., and Johnson, Scott P., 'Mental Rotation in Human Infants: A Sex Difference', *Psychological Science*, November 2008

Philip, R. C., Dauvermann, M. R., Whalley, H. C., Baynham, K., Lawrie, S. M., and Stanfield, A. C., 'A Systematic Review and Meta-analysis of the fMRI Investigation of Autism Spectrum Disorders', *Neuroscience and Biobehavioral Reviews*, February 2012

Pietschnig, J., Penke, L., Wicherts, J. M., Zeiler, M., and Voracek, M., 'Meta-analysis of Associations between Human Brain Volume and Intelligence Differences: How Strong are They and What Do They Mean?', *Neuroscience and Biobehavioral Reviews*, October 2015

Potegal, M., and Archer, J., 'Sex Differences in Childhood Anger and Aggression', *Child and Adolescent Psychiatric Clinics of North America*, July 2004

Prescott, C. A., Aggen, S. H., and Kendler, K. S., 'Sex-specific Genetic Influences on the Comorbidity of Alcoholism and Major Depression in a Population-based Sample of US Twins', *Archives of General Psychiatry*, August 2000

Quinn, P. C., and Liben, L. S., 'A Sex Difference in Mental Rotation in Young Infants', *Psychological Science*, November 2008

Raznahan, A., Shaw, P., Lalonde, F., Stockman, M., Wallace, G. L., et al., 'How Does Your Cortex Grow?', *Journal of Neuroscience*, May 2011

Raznahan, A., Toro, R., Daly, E., Robertson, D., Murphy, C., et al., 'Cortical Anatomy in Autism Spectrum Disorder: An in Vivo MRI Study on the Effect of Age', *Cerebral Cortex*, June 2010

Rucklidge, J. J., 'Gender Differences in Attention-Deficit/Hyperactivity Disorder', *Psychiatric Clinics of North America*, June 2010

Rynkiewicz, A., Schuller, B., Marchi, E., Piana, S., Camurri ,A., et al., 'An Investigation of the "Female Camouflage Effect" in Autism Using a Computerized ADOS-2 and a Test of Sex/Gender Differences', *Molecular Autism*, January 2016

Saha, S., Chant, D., Welham, J., and McGrath, J., 'A Systematic Review of the Prevalence of Schizophrenia', *PLoS Medicine*, May 2005

Samuel, D. B., and Widiger, T. A., 'Conscientiousness and Obsessive-Compulsive Personality Disorder', *Personality Disorders*, July 2011

Schmitt, David P., Realo, Anu, Voracek, Martin, and Allik, Jüri, 'Why Can't a Man Be More Like a Woman?: Sex Differences in Big Five Personality Traits across 55 Cultures', *Journal of Personality and Social Psychology*, January 2008

Substance Abuse and Mental Health Services Administration, *Results from the 2013 National Survey on Drug Use and Health: Summary of National Findings*, HHS Publication No. (SMA) 14-4863, NSDUH Series H-48, Substance Abuse and Mental Health Services Administration, 2014

Van Den Eeden, S. K., Tanner, C. M., Bernstein, A. L., Fross, R. D., Leimpeter, A., et al., 'Incidence of Parkinson's Disease: Variation by Age, Gender, and Race/Ethnicity', *American Journal of Epidemiology*, June 2003

Weissman, M. M., Bland, R. C., Canino, G. J., Faravelli, C., Greenwald, S., et al., 'Cross-national Epidemiology of Major Depression and Bipolar Disorder', *Journal of the American Medical Association*, July 1996

Winstanley, C. A., Eagle, D. M., and Robbins, T. W., 'Behavioral Models of Impulsivity in Relation to ADHD: Translation between Clinical and Preclinical Studies', *Clinical Psychology Review*, August 2006

Chapter 4

Bogen, J. E., and Bogen, G. M., 'Wernicke's Region: Where Is It?', *Annals of the New York Academy of Sciences*, October 1976

Carter, D. E., and Eckerman, D. A., 'Symbolic Matching by Pigeons: Rate of Learning Complex Discriminations Predicted from Simple Discriminations', *Science*, February 1975

Charles, S. T., and Carstensen, L. L., 'Social and Emotional Aging', *Annual Review of Psychology*, 2010

Colom, Roberto, Lluis-Font, Josep M., and Andrés-Pueyo, Antonio, 'The Generational Intelligence Gains Are Caused by Decreasing Variance in the Lower Half of the Distribution: Supporting Evidence for the Nutrition Hypothesis', *Intelligence*, 33, 2005

Démonet, J. F., Chollet, F., Ramsay, S., Cardebat, D., Nespoulous, J. L., et al., 'The Anatomy of Phonological and Semantic Processing in Normal Subjects', *Brain*, December 1992

Fjell, Anders M., Grydeland, Håkon, Krogsrud, Stine K., Amlien, Inge, Rohani, Darius A., et al., 'Development and Aging of Cortical Thickness Correspond to Genetic Organization Patterns', *Proceedings of the National Academy of Sciences of the United States of America*, September 2015

Fortenbaugh, F. C., DeGutis, J., Germine, L., Wilmer, J. B., Grosso, M., et al., 'Sustained Attention across the Life Span in a Sample of 10,000: Dissociating Ability and Strategy', *Psychological Science*, September 2015

Grodzinsky, Yosef, and Santi, Andrea, 'The Battle for Broca's Region', *Trends in Cognitive Sciences*, December 2008

Hartshorne, Joshua K., and Germine, Laura T., 'When Does Cognitive Functioning Peak?: The Asynchronous Rise and Fall of Different Cognitive Abilities across the Life Span', *Psychological Science*, April 2015

Hatton, T. J., and Bray, B. E., 'Long Run Trends in the Heights of European Men, 19th–20th Centuries', *Economics and Human Biology*, December 2010

Inhelder, Barbel, and Piaget, Jean, *The Early Growth of Logic in the Child: Classification and Seriation*, Routledge & Kegan Paul, 1964

Knecht, S., Dräger B., Deppe, M., Bobe, L., Lohmann, H., et al., 'Handedness and Hemispheric Language Dominance in Healthy Humans', *Brain*, December 2000

Libertus, Klaus, Joh, Amy S., and Work Needham, Amy, 'Motor Training at 3 Months Affects Object Exploration 12 Months Later', *Developmental Science*, 2015

Longevity Science Advisory Panel, 'Life Expectancy: Past and Future Variations by Gender in England and Wales', www.longevitypanel.co.uk/_files/life-expectancy-by-gender.pdf, 2012

Müller, U., Burman, J. T., and Hutchison, S. M., 'The Developmental Psychology of Jean Piaget: A Quinquagenary Retrospective', *Journal of Applied Developmental Psychology*, January 2013

Mustafa, N., Ahearn, T. S., Waiter, G. D., Murray, A. D., Whalley, L. J., and Staff, R. T., 'Brain Structural Complexity and Life Course Cognitive Change', *NeuroImage*, July 2012

Neisser, Ulric, 'Rising Scores on Intelligence Tests: Test Scores Are Certainly Going up All Over the World, but Whether Intelligence Itself Has Risen Remains Controversial', *American Scientist*, September–October 1997

Pinker, Stephen, *The Language Instinct: How the Mind Creates Language: The New Science of Language and Mind*, Penguin, 1995

Ridler, K., Veijola, J. M., Tanskanen, P., Miettunen, J., Chitnis, X., et al., 'Fronto-cerebellar Systems Are Associated with Infant Motor and Adult Executive Functions in Healthy Adults but Not in Schizophrenia', *Proceedings of the National Academy of Sciences of the United States of America*, October 2006

Stiles, J., and Jernigan, T. L., 'The Basics of Brain Development', *Neuropsychology Review*, December 2010

Whalley, Lawrence J., *Understanding Brain Aging and Dementia: A Life Course Approach*, Columbia University Press, 2015

Chapter 5

Anderson, M. V., and Rutherford, M. D., 'Cognitive Reorganization During Pregnancy and the Postpartum Period: An Evolutionary Perspective', *Evolutionary Psychology*, October 2012

Brennen, Tim, 'Seasonal Cognitive Rhythms Within the Arctic Circle: An Individual Differences Approach', *Journal of Environmental Psychology*, June 2001

Brennen, T., Martinussen, M., Hansen, B. O., and Hjemdal, O., 'Arctic Cognition: A Study of Cognitive Performance in Summer and Winter at 69°N', *Applied Cognitive Psychology*, 13, 1999

Buchanan, T. W., Laures-Gore, J. S., and Duff, M. C., 'Acute Stress Reduces Speech Fluency', *Biological Psychology*, March 2014

Chamberlain, S. R., Robbins, T. W., Winder-Rhodes, S., Müller, U., Sahakian, B.J., et al., 'Translational Approaches to Frontostriatal Dysfunction in Attention-Deficit/Hyperactivity Disorder Using a Computerized Neuropsychological Battery', *Biological Psychiatry*, June 2011

Cho, K., 'Chronic "Jet Lag" Produces Temporal Lobe Atrophy and Spatial Cognitive Deficits', *Nature Neuroscience*, June 2001

Christensen, H., Leach, L. S., and Mackinnon, A., 'Cognition in Pregnancy and Motherhood: Prospective Cohort Study', *British Journal of Psychiatry*, February 2010

Coren, Stanley, 'Daylight Savings Time and Traffic Accidents', *New England Journal of Medicine*, April 1996

Danziger, S., Levav, J., and Avnaim-Pesso, L., 'Extraneous Factors in Judicial Decisions', *Proceedings of the National Academy of Sciences of the United States of America*, April 2011

Davies, G., Welham, J., Chant, D., Torrey, E. F., and McGrath, J., 'A Systematic Review and Meta-analysis of Northern Hemisphere Season of Birth Studies in Schizophrenia', *Schizophrenia Bulletin*, January 2003

de Bruin, E. J., van Run, C., Staaks, J., and Meijer, A. M., 'Effects of Sleep Manipulation on Cognitive Functioning of Adolescents: A Systematic Review', *Sleep Medicine Reviews*, April 2017

Diekelmann, S., and Born, J., 'The Memory Function of Sleep', *Nature Reviews Neuroscience*, February 2010

Duan, S., Lv, Z., Fan, X., Wang, L., Han, F., et al., 'Vitamin D Status and the Risk of Multiple Sclerosis: A Systematic Review and Meta-analysis', *Neuroscience Letters*, June 2014

Galioto, R., and Spitznagel, M. B., 'The Effects of Breakfast and Breakfast Composition on Cognition in Adults', *Advances in Nutrition*, May 2016

Geoffroy, P. A., Bellivier, F., Scott, J., and Etain, B., 'Seasonality and Bipolar Disorder: A Systematic Review, from Admission Rates to Seasonality of Symptoms', *Journal of Affective Disorders*, October 2014

Haq, A., Svobodová, J., Imran, S., Stanford, C., and Razzaque, M. S., 'Vitamin D Deficiency: A Single Centre Analysis of Patients from 136 Countries', *Journal of Steroid Biochemistry and Molecular Biology*, November 2016

Hoekzema, E., Barba-Müller, E., Pozzobon, C., Picado, M., Lucco, F., et al., 'Pregnancy Leads to Long-lasting Changes in Human Brain Structure', *Nature Neuroscience*, February 2017

Hogan, Candice L., Mata, Jutta, and Carstensen, Laura L., 'Exercise Holds Immediate Benefits for Affect and Cognition in Younger and Older Adults', *Psychology and Aging*, June 2013

Hoyland, A., Dye, L., and Lawton, C. L., 'A Systematic Review of the Effect of Breakfast on the Cognitive Performance of Children and Adolescents', *Nutritional Research Reviews*, December 2009

Hwang, J., Brothers, R. M., Castelli, D. M., Glowacki, E. M., Chen, Y. T., et al., 'Acute High-Intensity Exercise-Induced Cognitive Enhancement and Brain-Derived Neurotrophic Factor in Young, Healthy Adults', *Neuroscience Letters*, September 2016

Kasper, S., Wehr, T. A., Bartko, J. J., Gaist, P. A., and Rosenthal, N. E., 'Epidemiological Findings of Seasonal Changes in Mood and Behavior: A Telephone Survey of Montgomery County, Maryland', *Archives of General Psychiatry*, September 1989

Lupien, S. J., Maheu, F., Tu, M., Fiocco, A., and Schramek, T. E., 'The Effects of Stress and Stress Hormones on Human Cognition: Implications for the Field of Brain and Cognition', *Brain and Cognition*, December 2007

Maguire, Eleanor A., Gadian, David G., Johnsrude, Ingrid S., Good, Catriona D., Ashburner, John, et al., 'Navigation-related Structural Change in the Hippocampi of Taxi Drivers', *Proceedings of the National Academy of Sciences of the United States of America*, April 2000

Marquié, J. C., Tucker, P., Folkard, S., Gentil, C., and Ansiau, D., 'Chronic Effects of Shift Work on Cognition: Findings from the VISAT Longitudinal Study', *Occupational and Environmental Medicine*, April 2015

Martens, Sander, and Wyble, Brad, 'The Attentional Blink: Past, Present, and Future of a Blind Spot in Perceptual Awareness', *Neuroscience and Biobehavioral Reviews*, May 2010

Mazahery, H., Camargo, C. A. Jr, Conlon, C., Beck, K. L., Kruger, M. C., and von Hurst, P. R., 'Vitamin D and Autism Spectrum Disorder: A Literature Review', *Nutrients*, April 2016

McGrath, J. J., Burne, T. H., Féron, F., Mackay-Sim, A., and Eyles, D. W., 'Developmental Vitamin D Deficiency and Risk of Schizophrenia: A 10-year Update', *Schizophrenia Bulletin*, November 2010

Meyer, C., Muto, V., Jaspar, M., Kussé, C., Lambot, E., et al., 'Seasonality in Human Cognitive Brain Responses', *Proceedings of the National Academy of Sciences of the United States of America*, March 2016

Miller, Alison L., Seifer, Ronald, Crossin, Rebecca, and Lebourgeois, Monique K., 'Toddler's Self-regulation Strategies in a Challenge Context are Nap-dependent', *Journal of Sleep Research*, June 2015
Miller, Michelle A., 'The Role of Sleep and Sleep Disorders in the Development, Diagnosis, and Management of Neurocognitive Disorders', *Frontiers in Neurology*, October 2013
Nowson, C. A., McGrath, J. J., Ebeling, P. R., Haikerwal, A., Daly, R. M., et al., 'Vitamin D and Health in Adults in Australia and New Zealand: A Position Statement', *Medical Journal of Australia*, June 2012
Pantelis, C., Barnes, T. R., Nelson, H. E., Tanner, S., Weatherley, L., et al., 'Frontal-striatal Cognitive Deficits in Patients with Chronic Schizophrenia', *Brain*, October 1997
Petković, Miodrag S., *Famous Puzzles of Great Mathematicians*, American Mathematical Society, 2009
Reeves, Adam, and Sperling, George, 'Attention Gating in Short-term Visual Memory', *Psychological Review*, 93, 1986
Shallice, T., 'Specific Impairments of Planning', Philosophical Transactions of the Royal Society, June 1982
Sherry, D. F., and MacDougall-Shackleton, S. A., 'Seasonal Change in the Avian Hippocampus', *Frontiers in Neuroendocrinology*, April 2015
Sundström Poromaa, Inger, and Gingnell, Malin, 'Menstrual Cycle Influence on Cognitive Function and Emotion Processing – from a Reproductive Perspective', *Frontiers in Neuroscience*, November 2014
Toffoletto, S., Lanzenberger, R., Gingnell, M., Sundström Poromaa, I., and Comasco, E., 'Emotional and Cognitive Functional Imaging of Estrogen and Progesterone Effects in the Female Human Brain: A Systematic Review', *Psychoneuroendocrinology*, December 2014
Warren, R. E., and Frier, B. M., 'Hypoglycaemia and Cognitive Function', *Diabetes, Obesity and Metabolism*, September 2005
Yerkes, Robert M., and Dodson, John D., 'The Relation of Strength of Stimulus to Rapidity of Habit-Formation', *Journal of Comparative Neurology and Psychology*, November 1908

Chapter 6

Ackermann, H., 'Cerebellar Contributions to Speech Production and Speech Perception: Psycholinguistic and Neurobiological Perspectives', *Trends in Neurosciences*, June 2008
Aleccia, JoNel, 'Taking out Half a Kid's Brain Can Be Best Option to Stop Seizures, Research Confirms', Today.com, August 2013
American Association of Neurological Surgeons, 'Gunshot Wound Head Trauma', http://www.aans.org/en/Patients/Neurosurgical-Conditions-and-Treatments/Gunshot-Wound-Head-Trauma, May 2015
Anderson, Vicki, Spencer-Smith, Megan, and Wood, Amanda, 'Do Children

Really Recover Better?: Neurobehavioural Plasticity after Early Brain Insult', *Brain*, August 2011
Barton, Robert A., and Venditti, Chris, 'Rapid Evolution of the Cerebellum in Humans and Other Great Apes', *Current Biology*, October 2014
Barrouquere, Brett, 'Defense: Death Row inmate Has No Frontal Lobe', *Lexington Herald Leader*, http://www.kentucky.com/news/local/crime/article44370489.html, August 2012
Bash, Dana, '"Stronger, better, tougher:" Giffords Improves, but She'll Never Be the Same', http://edition.cnn.com/2013/04/09/politics/giffords-health/, April 2013
BBC News, 'James Cracknell "Lucky to Be Alive" after US Bike Crash', http://www.bbc.co.uk/news/entertainment-arts-11411630, September 2010
Bennett, Hayley M., Mok, Hoi Ping, Gkrania-Klotsas, Effrossyni, Tsai, Isheng J., Stanley, Eleanor J., et al., 'The Genome of the Sparganosis Tapeworm *Spirometra Erinaceieuropaei* Isolated from the Biopsy of a Migrating Brain Lesion', *Genome Biology*, November 2014
Biography.com, 'Gabrielle Giffords', http://www.biography.com/people/gabrielle-giffords-20550593
Boatman, D., Freeman, J., Vining, E., Pulsifer, M., Miglioretti, D., et al., 'Language Recovery after Left Hemispherectomy in Children with Late-onset Seizures', *Annals of Neurology, October* 1999
Brodey, Sam, 'Missouri is About to Execute a Man Who's Missing Part of His Brain', Motherjones.com, March 2015
Callahan, Maureen, 'Cole Cohen: "Am I going crazy? What's wrong with me?"', News Corporation Australia, May 2015
Choi, Charles, 'Strange but True: When Half a Brain Is Better than a Whole One', *Scientific American*, May 2007
Cohen, Cole, http://us.macmillan.com/author/colecohen
Cohen, Cole, *Head Case: My Brain and Other Wonders*, Henry Holt & Co., 2015
DeNoon, Daniel J., 'Gabrielle Giffords' Brain Injury: FAQ', http://www.webmd.com/brain/news/20110109/gabrielle-giffords-brain-injury-faq, January 2011
Feuillet, Lionel, Dufour, Henry, Pelletier, Jean, 'Brain of a White-collar Worker', *The Lancet*, July 2007
Glickstein, Mitch, 'What Does the Cerebellum Really Do?', *Current Biology*, October 2007
Gupta, Sujata, 'Will Gabrielle Giffords Recover?', *New Scientist*, January 2011
Hamilton, Jon, 'A Man's Incomplete Brain Reveals Cerebellum's Role in Thought And Emotion', http://www.npr.org/sections/health-shots/2015/03/16/392789753/a-man-s-incomplete-brain-reveals-cerebellum-s-role-in-thought-and-emotion, March 2015
Healy, Melissa, 'Beyond the Bullet: Surviving a Shot to the Head Carries Host of Challenges', http://phys.org/news/2011-01-bullet-surviving-shot-host.html, January 2011

Hemispherectomy Foundation, The, 'Facts about Hemispherectomy', http://hemifoundation.homestead.com/facts.html

Hogan, M. J., Staff, R. T., Bunting, B. P., Murray, A. D., Ahearn, T. S., et al., 'Cerebellar Brain Volume Accounts for Variance in Cognitive Performance in Older Adults', *Cortex*, April 2011

Holloway, V., Gadian, D. G., Vargha-Khadem, F., Porter, D. A., Boyd, S. G., and Connelly, A., 'The Reorganization of Sensorimotor Function in Children after Hemispherectomy: A Functional MRI and Somatosensory Evoked Potential Study', *Brain*, December 2000

Hopegood, Rosie, 'James Cracknell on his Devastating Accident: "My Brain Injury Turned Me into a Completely Different Person"', Mirror.co.uk, October 2015

Johnson, Sara B., Blum, Robert W., and Giedd, J. N., 'Adolescent Maturity and the Brain: The Promise and Pitfalls of Neuroscience Research in Adolescent Health Policy', *Journal of Adolescent Health*, September 2009

Lew, Sean M., 'Hemispherectomy in the Treatment of Seizures: A Review', *Translational Pediatrics*, July 2014

Lin, Y., Harris, D. A., Curry, D. J., and Lam S., 'Trends in Outcomes, Complications, and Hospitalization Costs for Hemispherectomy in the United States for the Years 2000–2009', *Epilepsia*, January 2015

Macmillan, Malcolm, 'Phineas Gage – Unravelling the Myth', *The Psychologist*, https://thepsychologist.bps.org.uk/volume-21/edition-9/phineas-gage-unravelling-myth, September 2008

Marquez de la Plata, C. D., Hart, T., Hammond, F. M., Frol, A. B., Hudak, A., et al., 'Impact of Age on Long-term Recovery From Traumatic Brain Injury', *Archives of Physical Medicine and Rehabilitation*, May 2008

Marshall, Kelly, and Marrapodi, Eric, 'Born with Half a Brain, Woman Living Full Life', http://edition.cnn.com/2009/HEALTH/10/12/woman.brain/index.html?iref=24hours, October 2009

Monk, V., 'James Cracknell: I won't let doctors tell me what to do', Telegraph.co.uk, February 2015

Moosa, Ahsan N. V., Jehi, Lara, Marashly, Ahmad, Cosmo, Gary, Lachhwani, Deepak, et al.,. 'Long-term Functional Outcomes and Their Predictors after Hemispherectomy in 115 Children', *Epilepsia*, October 2013

Mosenthal, A. C., Livingston, D. H., Lavery, R. F., Knudson, M. M., Lee, S., et al., 'The Effect of Age on Functional Outcome in Mild Traumatic Brain Injury: 6-Month Report of a Prospective Multicenter Trial', *Journal of Trauma*, May 2004

Muckli, Lars, Naumer, Marcus J., and Singer, W., 'Bilateral Visual Field Maps in a Patient with Only One Hemisphere', *Proceedings of the National Academy of Sciences of the United States of America*, June 2009

National Institute of Neurological Disorders and Stroke, Agenesis of the Corpus Callosum information page, http://www.ninds.nih.gov/disorders/agenesis/agenesis.htm, May 2016

Nudo, Randolph J., 'Recovery after Brain Injury: Mechanisms and Principles', *Frontiers in Human Neuroscience*, December 2013

O'Driscoll, Kieran, and Leach, John Paul, '"No longer Gage": An Iron Bar Through the Head', *British Medical Journal*, December 1998

Paradiso, S., Andreasen, N. C., O'Leary, D. S., Arndt, S., and Robinson, R. G., 'Cerebellar Size and Cognition: Correlations with IQ, Verbal Memory and Motor Dexterity', *Neuropsychiatry, Neuropsychology and Behavioral Neurology*, January 1997

Pilkington, Ed, 'Missouri Executes Cecil Clayton, State's Oldest Death-row Inmate', https://www.theguardian.com/world/2015/mar/18/missouri-executes-cecil-clayton-supreme-court, March 2015

Schell-Apacik, C. C., Wagner, K., Bihler, M., Ertl-Wagner, B., Heinrich, U., et al., 'Agenesis and Dysgenesis of the Corpus Callosum: Clinical, Genetic and Neuroimaging Findings in a Series of 41 Patients', *American Journal of Medical Genetics*, October 2008

Teaches, Melyssa, 'Sharon Parker: The Woman with the Mysterious Brain', http://mymultiplesclerosis.co.uk/ep/sharon-parker-the-woman-with-the-mysterious-brain/, July 2015

Tovar-Moll, Fernanda, Monteiro, Myriam, Andrade, Juliana, Bramatia, Ivanei E., Vianna-Barbosa, Rodrigo, et al., 'Structural and Functional Brain Rewiring Clarifies Preserved Interhemispheric Transfer in Humans Born without the Corpus Callosum', *Proceedings of the National Academy of Sciences of the United States of America*, May 2014

University of Glasgow, 'Scientists Reveal Secret of Girl with "All Seeing Eye"', University of Glasgow website, July 2009

Yu, Feng, Jiang, Qing-jun, Sun, Xi-yan, and Zhang, Rong-wei, 'A New Case of Complete Primary Cerebellar Agenesis: Clinical and Imaging Findings in a Living Patient', *Brain*, June 2015

Chapter 7

Achiron, A., Measuring disability progression in multiple sclerosis. J Neurol (2006) 253: vi31

Alcohol Concern, 'Alcohol-Related Brain Damage: What Is It?' factsheet, 2016

Alcoholconcern.org.uk, 'Alcohol Statistics', https://www.alcoholconcern.org.uk/alcohol-statistics, August 2016

Alcohol Pharmacology Education Partnership, The, 'Module 2: The ABCs of Intoxication', https://sites.duke.edu/apep/module-2-the-abcs-of-intoxication/

Alzheimer's Association, 'Dementia with Lewy Bodies', http://www.alz.org/dementia/dementia-with-lewy-bodies-symptoms.asp

Alzheimer's Disease International, 'Dementia Statistics', www.alz.co.uk/research/statistics

Alzheimer's Society, 'What Is dementia?', https://www.alzheimers.org.uk/site/scripts/documents.php?categoryID=200360

Ibid., 'The Progression of Alzheimer's Disease and Other Dementias', https://www.alzheimers.org.uk/site/scripts/documents_info.php?documentID=133, April 2015

Ibid., 'What is Alcohol-related Brain Damage?'. https://www.alzheimers.org.uk/site/scripts/documents_info.php?documentID=98, October 2015

BBC News, 'Belfast Man with vCJD Dies after Long Battle', http://www.bbc.co.uk/news/uk-northern-ireland-12667709, March 2011

Ibid., 'First CJD Drug Trial Patient Dies', http://news.bbc.co.uk/1/hi/health/1687339.stm, 2 December 2001

Bloudoff-Indelicato, Mollie, 'Jack Osbourne: "Don't Let MS Control Your Life"', http://www.everydayhealth.com/multiple-sclerosis/living-with/jack-osbourne-dont-let-ms-control-your-life/, February 2016

Brockes, Emma, 'To the last breath', https://www.theguardian.com/education/2002/jan/15/medicalscience.health, 15 January 2002

Collie, D. A., Summers, D. M., Sellar, R. J., Ironside, J. W., Cooper, S., et al., 'Diagnosing Variant Creutzfeldt-Jakob Disease with the Pulvinar Sign: MR Imaging Findings in 86 Neuropathologically Confirmed Cases', *AJNR American Journal of Neuroradiology*, September 2003

Dailymail.co.uk, 'Fresh Hope as CJD Victim Improves', http://www.dailymail.co.uk/health/article-66176/Fresh-hope-CJD-victim-improves.html

Day, E., Bentham, P. W., Callaghan, R., Kuruvilla, T., and George, S., 'Thiamine for Prevention and Treatment of Wernicke-Korsakoff Syndrome in People Who Abuse Alcohol', *Cochrane Database of Systematic Reviews*, Issue 7, July 2013

De Stefano, N., Airas, L., Grigoriadis, N., Mattle, H. P., O'Riordan, J., et al. 'Clinical Relevance of Brain Volume Measures in Multiple Sclerosis', *CNS Drugs*, February 2014

European Multiple Sclerosis Platform, *Defeating MS Together: The European Code of Good Practice in MS*, September 2014

Harmon, Katherine, 'How Has Stephen Hawking Lived Past 70 with ALS?', *Scientific American*, January 2012

Hawking.org.uk, 'Brief Biography', http://www.hawking.org.uk/about-stephen.html

Hellerstein, David, 'Depression and Anxiety Disorders Damage Your Brain, Especially When Untreated', *Psychology Today*, https://www.psychologytoday.com/blog/heal-your-brain/201107/depression-and-anxiety-disorders-damage-your-brain-especially-when, July 2011

Help for Alzheimer's Families, 'Americans Rank Alzheimer's as Most Feared Disease', http://www.helpforalzheimersfamilies.com/alzheimers-dementia-care-services/alzheimers_feared_disease/, November 2012

Hendrick, Bill, 'Americans Worry about Getting Alzheimer's', http://www.webmd.com/alzheimers/news/20110223/americans-worry-about-getting-alzheimers, February 2011

Honig, L. S., and Mayeux, R., 'Natural History of Alzheimer's Disease', *Aging*, https://www.ncbi.nlm.nih.gov/pubmed/11442300, June 2001

Insel, Thomas, 'The Global Cost of Mental Illness', National Institute of Mental Health, https://www.nimh.nih.gov/about/director/2011/the-global-cost-of-mental-illness.shtml, September 2011

Kantarci, K., Lesnick, T., Ferman, T. J., Pryzbelski, S. A., Boeve, B. F., et al., 'Hippocampal Volumes Predict Risk of Dementia with Lewy Bodies in Mild Cognitive Impairment', *Neurology*, November 2016

Lillo, P., and Hodges, J. R., 'Cognition and Behaviour in Motor Neurone Disease', *Current Opinion in Neurology*, December 2010

Luerding, Ralf, Gebel, Sophie, Gebel, Eva-Maria, Schwab-Malek, Susanne, and Weissert, Robert, 'Influence of Formal Education on Cognitive Reserve in Patients with Multiple Sclerosis', *Frontiers in Neurology*, March 2016

McCoy, Terrence, 'How Stephen Hawking Is Still Alive, Defying ALS and the Worst Expectations', independent.co.uk, http://www.independent.co.uk/life-style/gadgets-and-tech/news/how-stephen-hawking-is-still-alive-defying-als-and-the-worst-expectations-10074974.html, 27 February 2015

Mental Health Foundation, *Fundamental Facts About Mental Health 2015*, https://www.mentalhealth.org.uk/publications/fundamental-facts-about-mental-health-2015, October 2015

Mezzapesa, D. M., Ceccarelli, A., Dicuonzo, F., Carella, A., De Caro, M. F., et al., 'Whole-Brain and Regional Brain Atrophy in Amyotrophic Lateral Sclerosis', *American Journal of Neuroradiology*, February 2007

Mirror.co.uk., 'Longest Surviving Victim of vCJD Holly Mills Dies in Her Sleep', http://www.mirror.co.uk/news/technology-science/longest-surviving-victim-of-vcjd-holly-98217, 27 November 2011

Motor Neurone Disease Association, 'Different Types of MND', http://www.mndassociation.org/what-is-mnd/different-types-of-mnd/

MS International Federation, 'What Is MS?', https://www.msif.org/about-ms/what-is-ms/, October 2016

National CJD Research & Surveillance Unit, The, http://www.cjd.ed.ac.uk/index.html, University of Edinburgh website

National Institute on Aging, 'Alzheimer's Disease: Unraveling the Mystery – The Changing Brain in Healthy Aging', https://www.nia.nih.gov/alzheimers/publication/part-1-basics-healthy-brain/changing-brain-healthy-aging, January 2015

National Institute on Alcohol Abuse and Alcoholism, *The Neurotoxicity of Alcohol*, http://pubs.niaaa.nih.gov/publications/10report/chap02e.pdf,

Ibid., 'Alcohol Alert', no. 46, http://pubs.niaaa.nih.gov/publications/aa46.htm, December 1999

National Institute of Neurological Disorders and Stroke, 'Creutzfeldt-Jakob Disease Fact Sheet', http://www.ninds.nih.gov/disorders/cjd/detail_cjd.htm, March 2003

NHS Choices, 'Creutzfeldt-Jakob Disease', http://www.nhs.uk/conditions/Creutzfeldt-Jakob-disease/Pages/Introduction.aspx, July 2015

Ibid., 'Multiple Sclerosis – Symptoms', http://www.nhs.uk/Conditions/Multiple-sclerosis/Pages/Symptoms.aspx

Office for National Statistics, 'Deaths Registered in England and Wales (Series DR): 2015', https://www.ons.gov.uk/peoplepopulationandcommunity/birthsdeathsandmarriages/deaths/bulletins/deathsregisteredinenglandandwalesseriesdr/2015

Oliver, Joe, 'CJD Survivor Still Defying the Odds', belfasttelegraph.co.uk, http://www.belfasttelegraph.co.uk/sunday-life/cjd-survivor-still-defying-the-odds-28459301.html, December 2008

Parry, A., Baker, I., Stacey, R., and Wimalaratna. S., 'Long term Survival in a Patient with Variant Creutzfeldt–Jakob Disease Treated with Intraventricular Pentosan Polysulphate', *Journal of Neurology, Neurosurgery & Psychiatry*, July 2007

Patients Association, The, 'Dementia Overtakes Cancer as UK's Most Feared Illness', http://www.patients-association.org.uk/press-release/dementia-overtakes-cancer-uks-feared-illness/, February 2015

Peters, R., 'Ageing and the Brain', *Postgraduate Medical Journal*, February 2006

ScienMag.com, 'Study: Lack of Brain Shrinkage May Help Predict Who Develops Dementia with Lewy Bodies', . http://scienmag.com/study-lack-of-brain-shrinkage-may-help-predict-who-develops-dementia-with-lewy-bodies/, November 2016

Shiee, N., Bazin, P. L., Zackowski, K. M., Farrell, S. K., Harrison, D. M., et al., 'Revisiting Brain Atrophy and Its Relationship to Disability in Multiple Sclerosis', *PLoS One*, May 2012

Stanford Medicine News Center, 'Different Mental Disorders Linked to Same Brain-matter Loss, Study Finds', https://med.stanford.edu/news/all-news/2015/02/different-mental-disorders-cause-same-brain-matter-loss.html, 4 February 2015

Steinman, Lawrence, 'No Quiet Surrender: Molecular Guardians in Multiple Sclerosis Brain', *Journal of Clinical Investigation*, April 2015

Stern, Y., 'Cognitive Reserve in Ageing and Alzheimer's Disease', *Lancet Neurology*, November 2012

Stern, Yaakov, 'Cognitive Reserve and Alzheimer Disease', *Alzheimer Disease & Associated Disorders*, Vol. 20, April/June 2006

Sullivan, Edith V., Harris, R. Adron, and Pfefferbaum, Adolf, 'Alcohol's Effects on Brain and Behavior', *Alcohol Research & Health*, January 2010

Telegraph.co.uk, 'Ozzy Osbourne's Son Jack Diagnosed with Multiple Sclerosis', http://www.telegraph.co.uk/culture/music/music-news/9337002/Ozzy-Osbournes-son-Jack-diagnosed-with-multiple-sclerosis.html, June 2012

Topiwala, Anya, Allan, Charlotte L., Valkanova, Vyara, Zsoldos, Enikő, Filippini, Nicola, et al., 'Moderate Alcohol Consumption as Risk Factor for Adverse Brain Outcomes and Cognitive Decline: Longitudinal Cohort Study', *British Medical Journal*, May 2017

UCSF Memory and Aging Center, 'Alzheimer's Disease', http://memory.ucsf.edu/education/diseases/alzheimer
U.S. Department of Health and Human Services, *10th Special Report to the U.S. Congress on Alcohol and Health: Highlights from Current Research*, June 2000
World Health Organization, 'Variant Creutzfeldt-Jakob Disease', http://www.who.int/mediacentre/factsheets/fs180/en/, February 2012
YouGov UK, 'Cancer Britons Most Feared Disease', https://yougov.co.uk/news/2011/08/15/cancer-britons-most-feared-disease/, August 2011

Chapter 8

Barker, D., and Osmond, C.. 'Infant Mortality, Childhood Nutrition, and Ischaemic Heart Disease in England and Wales', *The Lancet*, May 1986
Barnett, J. H., Salmond, C. H., Jones, P. B., and Sahakian, B. J., 'Cognitive Reserve in Neuropsychiatry', *Psychological Medicine*, August 2006
Barnett, Jennifer H., Hachinski, Vladimir, and Blackwell, Andrew D., 'Cognitive Health Begins at Conception: Addressing Dementia as a Lifelong and Preventable Condition, *BMC Medicine*, November 2013
Batouli, S. A., Trollor, J. N., Wen, W., and Sachdev, P. S., 'The Heritability of Volumes of Brain Structures and Its Relationship to Age: A Review of Twin and Family Studies', *Ageing Research Reviews*, January 2014
Belsky, D. W., Caspi, A., Israel, S., Blumenthal, J. A., Poulton, R., and Moffitt, T. E. 'Cardiorespiratory Fitness and Cognitive Function in Midlife: Neuroprotection or Neuroselection?', Annals of Neurology, April 2015
Black, R. E., Victora, C. G., Walker, S. P., Bhutta, Z. A., Christian, P., et al., 'Maternal and Child Undernutrition and Overweight in Low-income and Middle-income Countries', *The Lancet*, August 2013
Bouchard, T. J., 'The Wilson Effect: The Increase in Heritability of IQ with Age', *Twin Research and Human Genetics*, October 2013
Bouchard, T. J. Jr, and McGue, M., 'Genetic and Environmental Influences on Human Psychological Differences', *Journal of Neurobiology*, January 2003
Cox, E. P., O'Dwyer, N., Cook, R., Vetter, M., Cheng, H. L., et al., 'Relationship between Physical Activity and Cognitive Function in Apparently Healthy Young to Middle-aged Adults: A Systematic Review', *Journal of Science and Medicine in Sport*, August 2016
Davies, G., Marioni, R. E., Liewald, D. C., Hill, W. D., Hagenaars, S. P., et al., 'Genome-wide Association Study of Cognitive Functions and Educational Attainment in UK Biobank (N=112 151)', *Molecular Psychiatry*, June 2016
Deary, I. J., Johnson, W., and Houlihan, L. M., 'Genetic Foundations of Human Intelligence', *Human Genetics*, July 2009
Ferguson, Christopher J., 'Do Angry Birds Make for Angry Children?: A Meta-analysis of Video Game Influences on Children's and Adolescents' Aggression, Mental Health, Pro-social Behavior, and Academic Performance', *Perspectives on Psychological Science*, September 2015

Gefen ,T., Peterson, M., Papastefan, S. T., Martersteck, A., Whitney, K., et al., 'Morphometric and Histologic Substrates of Cingulate Integrity in Elders with Exceptional Memory Capacity', *Journal of Neuroscience*, January 2015

Gillman, M. W., and Rich-Edwards, J. W., 'The Fetal Origin of Adult Disease: From Sceptic to Convert', *Paediatric and Perinatal Epidemiology*, July 2000

Goldman, A. S., 'The Immune System of Human Milk: Antimicrobial, Antiinflammatory and Immunomodulating Properties', *Pediatric Infectious Disease Journal*, August 1993

Hagenaars, S. P., Harris, S. E., Davies, G., Hill, W. D., Liewald, D. C., et al., 'Shared Genetic Aetiology between Cognitive Functions and Physical and Mental Health in UK Biobank (N=112 151) and 24 GWAS Consortia', *Molecular Psychiatry*, November 2016

Hales, C. N., and Barker, D. J., 'The Thrifty Phenotype Hypothesis', *British Medical Bulletin*, 2001

Harris, Judith Rich, *The Nurture Assumption: Why Children Turn Out the Way They Do*, Bloomsbury, 1998

Harrison, Theresa M., Weintraub, Sandra, Mesulam, M.-Marsel, and Rogalski, Emily, 'Superior Memory and Higher Cortical Volumes in Unusually Successful Cognitive Aging', *Journal of the International Neuropsychological Society*, November 2012

Hopkins, M. E., Davis, F. C., Vantieghem, M. R., Whalen, P. J., and Bucci, D. J., 'Differential Effects of Acute and Regular Physical Exercise on Cognition and Affect', *Neuroscience*, July 2012

Horta, Bernardo L., and Victora, Cesar G., *Short-term Effects of Breastfeeding: A Systematic Review on the Benefits of Breastfeeding on Diarrhoea and Pneumonia Mortality*, World Health Organization Institutional Repostiory for Information Sharing, 2013

Horta, B. L., Loret de Mola, C., and Victora, C. G., 'Breastfeeding and Intelligence: A Systematic Review and Meta-analysis', *Acta Paediatrica*, December 2015

Kormos, C. E., Wilkinson, A. J., Davey, C. J., and Cunningham, A. J., 'Low Birth Weight and Intelligence in Adolescence and Early Adulthood: A Meta-analysis', *Journal of Public Health*, June 2014

Kramer, M. S., 'Determinants of Low Birth Weight: Methodological Assessment and Meta-analysis', *Bulletin of the World Health Organization*, 1987

Kramer, M. S., Aboud, F., Mironova, E., Vanilovich, I., Platt, R. W., et al., 'Breastfeeding and Child Cognitive Development: New Evidence from a Large Randomized Trial', *Archives of General Psychiatry*, May 2008

Ksir, C., and Hart, C. L., 'Cannabis and Psychosis: A Critical Overview of the Relationship', *Current Psychiatry Reports*, February 2016

Lees, C., and Hopkins, J., 'Effect of Aerobic Exercise on Cognition, Academic Achievement, and Psychosocial Function in Children: A Systematic Review of Randomized Control Trials', *Preventing Chronic Disease*, October 2013

Loret de Mola, C., de França, G. V., Quevedo, Lde A., and Horta, B. L., 'Low Birth Weight, Preterm Birth and Small for Gestational Age Association with Adult Depression: Systematic Review and Meta-analysis', *British Journal of Psychiatry*, November 2014

Ma, Y., Goins, K. V., Pbert, L., and Ockene, J. K., 'Predictors of Smoking Cessation in Pregnancy and Maintenance Postpartum in Low-income Women', *Maternal and Child Health Journal*, December 2005

Marconi, A., Di Forti, M., Lewis, C. M., Murray, R. M., and Vassos, E., 'Meta-analysis of the Association Between the Level of Cannabis Use and Risk of Psychosis', *Schizophrenia Bulletin*, September 2016

Moffitt, T. E., Arseneault, L., Belsky, D., Dickson, N., Hancox, R. J., et al., 'A Gradient of Childhood Self-control Predicts Health, Wealth, and Public Safety', *Proceedings of the National Academy of Sciences of the United States of America*, February 2011

Moon, H. Y., Becke, A., Berron, D., Becker, B., Sah, N., et al., 'Running-Induced Systemic Cathepsin B Secretion Is Associated with Memory Function', *Cell Metabolism*, August 2016

Paul, Annie Murphy, *Origins : How the Nine Months before Birth Shape the Rest of our Lives*, New York: Free Press, 2011

Podewils, L. J., Guallar, E., Kuller, L. H., Fried, L. P., Lopez, O. L., et al., 'Physical activity, APOE Genotype, and Dementia Risk: Findings from the Cardiovascular Health Cognition Study', *American Journal of Epidemiology*, April 2005

Polderman, T. J., Benyamin, B., de Leeuw, C. A., Sullivan, P. F., van Bochoven, A., et al., 'Meta-analysis of the Heritability of Human Traits Based on Fifty Years of Twin Studies', *Nature Genetics*, July 2015

Raikkonen, K., Kajantie, E., Pesonen, A. K., Heinonen, K., Alastalo, H., et al., 'Early Life Origins Cognitive Decline: Findings in Elderly Men in the Helsinki Birth Cohort Study', *PLoS One*, 2013

Roig, M., Nordbrandt, S., Geertsen, S. S., and Nielsen, J. B., 'The Effects of Cardiovascular Exercise on Human Memory: A Review with Meta-analysis', *Neuroscience and Biobehavioral Reviews*, September 2013

Scarr, S., and McCartney, K., 'How People Make Their Own Environments: A Theory of Genotype Greater than Environment Effects', Child Development, April 1983

Schaefer, J. D., Caspi, A., Belsky, D. W., Harrington, H., Houts, R., et al., 'Enduring Mental Health: Prevalence and Prediction', Journal of Abnormal Psychology, February 2017

Slutske, W. S., Moffitt, T. E., Poulton, R., and Caspi, A., 'Undercontrolled Temperament at Age 3 Predicts Disordered Gambling at Age 32: A Longitudinal Study of a Complete Birth Cohort', *Psychological Science*, May 2012

Spalding, K. L., Bergmann, O., Alkass, K., Bernard, S., Salehpour, M., et al., 'Dynamics of Hippocampal Neurogenesis in Adult Humans', *Cell*, June 2013
Stern, Y., 'Cognitive Reserve in Ageing and Alzheimer's Disease', *Lancet Neurology*, November 2012
Sun, F. W., Stepanovic, M. R., Andreano, J., Barrett, L. F., Touroutoglou, A., and Dickerson, B. C., 'Youthful Brains in Older Adults: Preserved Neuroanatomy in the Default Mode and Salience Networks Contributes to Youthful Memory in Superaging', *Journal of Neuroscience*, September 2016
van Oijen, M., de Jong, F. J., Witteman, J. C., Hofman, A., Koudstaal, P. J., and Breteler, M. M., 'Atherosclerosis and Risk for Dementia', *Annals of Neurology*, May 2007
van Praag, H., Shubert, T., Zhao, C., and Gage, F. H., 'Exercise Enhances Learning and Hippocampal Neurogenesis in Aged Mice', *Journal of Neuroscience*, September 2005
Woodby, L. L., Windsor, R. A., Snyder, S. W., Kohler, C. L., and Diclemente, C. C., 'Predictors of Smoking Cessation during Pregnancy', *Addiction*, February 1999

Chapter 9

Abbott, C. C., Gallegos, P., Rediske, N., Lemke, N. T., and Quinn, D. K., 'A Review of Longitudinal Electroconvulsive Therapy: Neuroimaging Investigations', *Journal of Geriatric Psychiatry and Neurology*, March 2014
Andersen, R. A., Kellis, S., Klaes, C., and Aflalo, T., 'Toward More Versatile and Intuitive Cortical Brain–Machine Interfaces', *Current Biology*, September 2014
BBC News, 'Paralysed Man Feeds Himself with Help of Implants', http://www.bbc.co.uk/news/health-39416974, 29 March 2017
Bharatbook.com, 'Global Cosmetic Surgery and Service Market Report 2015–2019', https://www.bharatbook.com/healthcare-market-research-reports-643332/global-cosmetic-surgery-service.html, March 2015
Biddle, S. J., Gorely, T., Marshall, S. J., Murdey, I., and Cameron, N., 'Physical Activity and Sedentary Behaviours in Youth: Issues and Controversies', *Journal of the Royal Society for the Promotion of Health*, January 2004
Bisagno, V., González, B., and Urbano, F. J., 'Cognitive Enhancers versus Addictive Psychostimulants: The Good and Bad Side of Dopamine on Prefrontal Cortical Circuits', *Pharmacological Research*, July 2016
Chen, H., Kwong, J. C., Copes, R., Tu, K., Villeneuve, P. J., et al., 'Living Near Major Roads and the Incidence of Dementia, Parkinson's Disease, and Multiple Sclerosis: A Population-based Cohort Study', *The Lancet*, February 2017
Darpa.mil, 'Neurotechnology Provides Near-Natural Sense of Touch', http://www.darpa.mil/news-events/2015-09-11, September 2015

Deer, T. R., Krames, E., Mekhail, N., Pope, J., Leong, M., et al., 'The Appropriate Use of Neurostimulation: New and Evolving Neurostimulation Therapies and Applicable Treatment for Chronic Pain and Selected Disease States. Neuromodulation Appropriateness Consensus Committee', *Neuromodulation*, August 2014

Dierckx, B., Heijnen, W. T., van den Broek, W. W., and Birkenhäger, T. K., 'Efficacy of Electroconvulsive Therapy in Bipolar versus Unipolar Major Depression: A Meta-analysis', *Bipolar Disorders*, March 2012

Eapen, B. C., Murphy, D. P., and Cifu, D. X., 'Neuroprosthetics in Amputee and Brain Injury Rehabilitation', *Experimental Neurology*, January 2017

Ernst & Young, *Seeking Sustainable Growth: The Luxury and Cosmetics Financial Factbook*, http://www.ey.com/Publication/vwLUAssets/EY_Factbook_2015/$FILE/EY-Factbook-2015.PDF, 2015

Etchells, Pete, Fletcher-Watson, Sue, Blakemore, Sarah-Jayne, Chambers, Chris, Kardefelt-Winther, Daniel, et al., 'Screen Time Guidelines Need to Be Built on Evidence, Not Hype', https://www.theguardian.com/science/head-quarters/2017/jan/06/screen-time-guidelines-need-to-be-built-on-evidence-not-hype, 6 January 2017

Federici, M., Latagliata, E. C., Rizzo, F. R., Ledonne, A., Gu, H. H., et al., 'Electrophysiological and Amperometric Evidence that Modafinil Blocks the Dopamine Uptake Transporter to Induce Behavioral Activation', *Neuroscience*, November 2013

George, Madeleine J., and Odgers, Candice L., 'Seven Fears and the Science of How Mobile Technologies May Be Influencing Adolescents in the Digital Age', *Perspectives on Psychological Science*, November 2015

Godinho, B. M., Malhotra, M., O'Driscoll, C. M., and Cryan, J. F., 'Delivering a Disease-modifying Treatment for Huntington's Disease', *Drug Discovery Today*, January 2015

Hawking.com, 'My Computer', http://www.hawking.org.uk/the-computer.html

Haz-map.com, 'In Post-Industrial Countries, What Is the Current Status of Our Environment Compared to 25 Years Ago?', http://www.haz-map.com/pollutio.htm, April 2011

Horvath, J. C., Forte, J. D., and Carter, O., 'Quantitative Review Finds No Evidence of Cognitive Effects in Healthy Populations From Single-session Transcranial Direct Current Stimulation (tDCS)', *Brain Stimulation*, May–June 2015

Iaccarino, H. F., Singer, A. C., Martorell, A J., Rudenko, A., Gao, F., et al., 'Gamma Frequency Entrainment Attenuates Amyloid Load and Modifies Microglia', *Nature*, December 2016

Jarvis, S., and Schultz, S. R., 'Prospects for Optogenetic Augmentation of Brain Function', *Frontiers in Systems Neuroscience*, November 2015

Kalia, L. V., Kalia, S. K., and Lang, A. E., 'Disease-modifying Strategies for Parkinson's Disease', *Movement Disorders*, September 2015

Kirik, D., Cederfjäll, E., Halliday, G., and Petersén, A., 'Gene Therapy for Parkinson's Disease: Disease Modification by GDNF Family of Ligands', *Neurobiology of Disease*, January 2017

Lefaucheur, J. P., André-Obadia, N., Antal, A., Ayache, S. S., Baeken, C., et al., 'Evidence-based Guidelines on the Therapeutic Use of Repetitive Transcranial Magnetic Stimulation (rTMS)', Clinical Neurophysiology, November 2014

Lewis, P. M., Ackland, H. M., Lowery, A. J., and Rosenfeld, J. V., 'Restoration of Vision in Blind Individuals Using Bionic Devices: A Review with a Focus on Cortical Visual Prostheses', *Brain Research*, January 2015

LeWitt, P. A., Rezai, A. R., Leehey, M. A., Ojemann, S. G., Flaherty, A. W., et al., 'AAV2-GAD Gene Therapy for Advanced Parkinson's Disease: A Double-blind, Sham-surgery Controlled, Randomised Trial', *Lancet Neurology*, April 2011

Miocinovic, S., Somayajula, S., Chitnis, S., and Vitek, J. L., 'History, Applications, and Mechanisms of Deep Brain Stimulation', *JAMA Neurology*, February 2013

Mitteroecker, P., Huttegger, S. M., Fischer, B., and Pavlicev, M., 'Cliff-edge Model of Obstetric Selection in Humans', *Proceedings of the National Academy of Sciences of the United States of America*, December 2016

Mohammadi, Dara, 'Huntington's Disease: The New Gene Therapy That Patients Cannot Afford', https://www.theguardian.com/science/2016/may/15/huntingtons-disease-drugs-cure-research-poor-families-colombia-corporate-responsibility, 15 May 2016

Niparko, J. K., Tobey, E. A., Thal, D. J., Eisenberg, L. S., Wang, N Y., et al., 'Spoken Language Development in Children Following Cochlear Implantation', *Journal of the American Medical Association*, April 2010

Plasticsurgery.org, 'Plastic Surgery Statistics Show New Consumer Trends', https://www.plasticsurgery.org/news/press-releases/plastic-surgery-statistics-show-new-consumer-trends, 26 February 2015

Ramaswamy, S., and Kordower, J. H., 'Gene Therapy for Huntington's Disease', *Neurobiology of Disease*, November 2012

Repantis, D., Laisney, O., and Heuser, I., 'Acetylcholinesterase Inhibitors and Memantine for Neuroenhancement in Healthy Individuals: A Systematic Review', *Pharmacological Research*, June 2010

Repantis, D., Schlattmann, P., Laisney, O., and Heuser, I., 'Modafinil and Methylphenidate for Neuroenhancement in Healthy Individuals: A Systematic Review', *Pharmacological Research*, September 2010

Roy, D. S., Arons, A., Mitchell, T. I., Pignatelli, M., Ryan, T. J., and Tonegawa, S., 'Memory Retrieval by Activating Engram Cells in Mouse Models of Early Alzheimer's Disease', *Nature*, March 2016

Sahakian, B. J., Bruhl, A. B., Cook, J., Killikelly, C., Savulich, G., et al., 'The Impact of Neuroscience on Society: Cognitive Enhancement in

Neuropsychiatric Disorders and in Healthy People', *Philosophical Transactions of the Royal Society of London, Biological Sciences*, September 2015
Shin, J. W., Kim, K. H., Chao, M. J., Atwal, R. S., Gillis, T., et al., 'Permanent Inactivation of Huntington's Disease Mutation by Personalized llele-specific CRISPR/Cas9', *Human Molecular Genetics*, October 2016
Slotema, C. W., Blom, J. D., Hoek, H. W., and Sommer, I. E., 'Should We Expand the Toolbox of Psychiatric Treatment Methods to Include Repetitive Transcranial Magnetic Stimulation (rTMS)?: A Meta-analysis of the Efficacy of rTMS in Psychiatric Disorders', *Journal of Clinical Psychiatry*, July 2010
Smith, M. E., and Farah, M. J., 'Are Prescription Stimulants "smart pills"?: The Epidemiology and Cognitive Neuroscience of Prescription Stimulant Use by Normal Healthy Individuals', *Psychologial Bulletin*, September 2011
Sparreboom, M., van Schoonhoven, J., van Zanten, B. G., Scholten, R. J., Mylanus, E. A., et al., 'The Effectiveness of Bilateral Cochlear Implants for Severe-to-Profound Deafness in Children: A Systematic Review', *Otology & Neurotology*, September 2010
UCL Huntington's Disease Research, 'Trial of Innovative Drug, Developed by Ionis Pharmaceuticals, Aims to Reduce Production of the Toxic Protein that Causes Devastating Brain Disease', http://hdresearch.ucl.ac.uk/2015/10/first-patients-treated-with-gene-silencing-drug-isis-httrx-for-huntingtons-disease-2/, October 2015
van der Lely, S., Frey, S., Garbazza, C., Wirz-Justice, A., Jenni, O. G., et al, 'Blue Blocker Glasses as a Countermeasure for Alerting Effects of Evening Light-emitting Diode Screen Exposure in Male Teenagers', *Journal of Adolescent Health*, January 2015
van Schoonhoven, J., Sparreboom, M., van Zanten, B. G., Scholten, R. J., Mylanus, E. A., et al., 'The Effectiveness of Bilateral Cochlear Implants for Severe-to-Profound Deafness in Adults: A Systematic Review', *Otology & Neurotology*, February 2013
Vastag, B., 'Poised to Challenge Need for Sleep, "Wakefulness Enhancer" Rouses Concerns', *Journal of the American Medical Association*, January 2004
Xie, L., Kang, H., Xu, Q., Chen, M. J., Liao, Y., et al., 'Sleep Drives Metabolite Clearance from the Adult Brain', *Science*, October 2013

Index

Page numbers in *italics* refer to figures